FAMILY GOAT-KEEPING

by

W. O'CONNELL HOLMES

(Editor of " GOATS " Magazine).

British Library Cataloguing-in-Publication Data
A catalogue record for this book is available from the
British Library

Goat Farming

The domestic goat (*Capra aegagrus hircus*) is a subspecies of goat domesticated from the wild goat of southwest Asia and Eastern Europe. The goat is a member of the family Bovidae and there are now over 300 distinct breeds. Goats are one of the oldest domesticated species, and have been used for their milk, meat, hair, and skins over much of the world – their relationship with humans has an incredibly long, useful and varied history. The most recent genetic analysis confirms the archaeological evidence that the wild Bezoar ibex of the Zagros Mountains are the likely origin of almost all domestic goats today.

Neolithic farmers began to herd wild goats for easy access to milk and meat, primarily, as well as for their dung, which was used as fuel, and their bones, hair, and sinew for clothing, building, and tools. The earliest remnants of domesticated goats have been found in Iran, dating at around 10,000 years ago, and remains have also been found at archaeological sites in Jericho, Choga Mami Djeitun and Çayönü, dating the domestication of goats in Western Asia at between 8000 and 9000 years ago. Today (as then), Goat farming is a profitable business, due to its relatively low investments - and multi functional utility. A goat is useful to humans when it is living and when it is dead, first as a renewable provider of milk, manure, and fibre, and then as meat and hide. Some charities even provide goats to impoverished people in poor countries, because goats are far easier and cheaper to manage than cattle – but provide just as much value.

The particular housing used for goats depends on the intended use of the animal, but also on the region of the world in which it is raised. Historically, domestic goats were generally kept in herds that wandered on hills or other grazing areas, often tended by goatherds who were frequently children or adolescents, similar to the more widely known shepherd; a method still used in the present day. Most commonly though, goats are kept for their dairy produce (especially in Asia and Nepal), which means they are far more likely to be kept in barns, close to the farm. Dairy goats are generally pastured in summer and may be stabled during the winter, whereas meat goats are more frequently pastured year-round, and may be kept many miles from their barns.

Aside from their uses for meat, dairy as well as their dung and hide, goats have been used by humans to clear unwanted vegetation for centuries. They have been described as 'eating machines' and 'biological control agents', and have proved incredibly successful in such endeavours. There has been a resurgence of this practice in North America since 1990, when herds were used to clear dry brush from California hillsides thought to be endangered by potential wildfires. Since then, numerous public and private agencies have hired private herds to perform similar tasks. This practice has also become popular in the Pacific Northwest, where they are used to remove invasive species not easily removed by humans, including (thorned) blackberry vines and poison oak. This wonderful and diverse species continues to be of much use to the human race, and it is for this reason that goat farming retains its interest.

With a sound and comfortable shed, a well-fenced exercise yard, facilities for growing a few vegetables, and occasional opportunities for gathering food from hedgerows, goat keeping is possible for thousands of people, even those who are at work all day, who hitherto have considered that goats can be kept only where much space is available. Here is a little herd kept on the stall-feeding, or intensive system, whose semi-restricted life in no

HE AUTHOR gratefully acknowledges the unstinting help and advice given him in the preparation of this book by Mrs. W. G. Lucock, Mrs. R. Richens, Miss N. M. B. Smales, Messrs. G. Wyatt, E. T. Harding, T. P. Slipper, H. V. Cook, and many others behind the scenes.

Foreword.

I never expected quite so soon to be writing a foreword to the second issue of " Family Goat Keeping." The first edition has met with an encouraging reception, and many letters of praise and thanks have been received from readers who have found the little book helpful. In turn, I thank all friends who have worked hard to make the book known.

It must be mentioned that basic rations have been withdrawn for the time being, and while this apparently increases feeding difficulties, the position is not as bad as it seems. There is an abundance of natural food available which the right type of goat will convert into a reasonably good milk supply. And whenever the semi-stall-fed goat cannot go out to graze and browse then it is up to the owner to go out and do the grazing and browsing ! Every effort made to use every scrap of suitable greenstuff for conversion into milk is another step towards winning the war.

The demand for goats continues to grow. If at first the goat you do obtain is not quite what you would like, be patient. Breed from her, using good males, and it will not be long before you have good milkers.

Finally, I shall be pleased to help anyone with a goat problem, on receipt of return postage.

W. O'CONNELL HOLMES.

Colchester,

FAMILY GOAT-KEEPING.

CONTENTS.

CHAPTER		PAGE
I	INTRODUCTION	1
II	GETTING YOUR GOAT	3
III	YOUR GOAT'S HOUSE AND RUN	8
IV	FEEDING YOUR GOAT	14
V	YOUR GUIDE TO BREEDING	24
VI	KIDDING WITHOUT WORRY	27
VII	THE KIDS	30
VIII	DISBUDDING KIDS	35
IX	MILK-oI	37
X	MAKING YOUR OWN BUTTER AND CHEESE	39
XI	GOAT MEAT FOR THE TABLE	43
XII	CURING KID SKINS	45
XIII	AILMENTS	47
XIV	IN GENERAL	52
XV	A FEW RECIPES	56
XVI	A FAMILY MAN STARTS GOAT-KEEPING	58

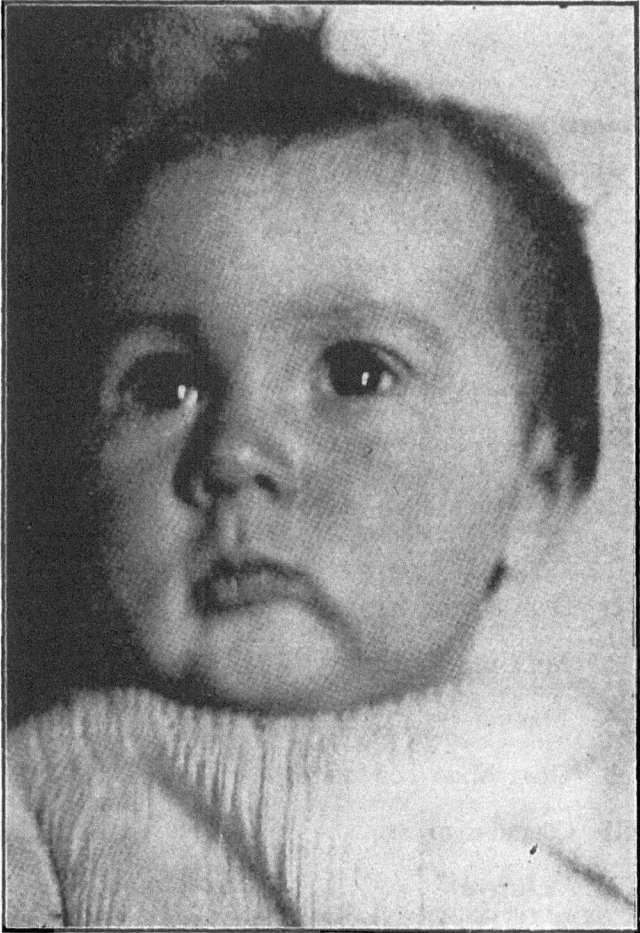

A GOATS' MILK BABY.

One of the author's little "herd," living proof of the value of goats' milk as a baby food.

CHAPTER I.

Introduction.

THIS war, like the last, has brought an urgent demand for goats, to provide the family milk supply. The difference is that, in the interim, there has been something of a revolution in goat-keeping. New breeds have appeared, old breeds have been improved almost beyond recognition, milk yields of individuals have risen to staggering proportions, and the average yield from goats has greatly increased. Further, new methods and ideas concerning the keeping of goats have been introduced and proved.

True, one still occasionally meets with prejudice concerning the " smell " of goats and the " taste " of goats' milk, but we can laugh at such fast-receding ignorance, supported all round, as we are, by evidence to the contrary, supplied by real practical honest-to-goodness people from the poorest cottagers to the highest medical authorities. Evidence which not only proves the purity of goats' milk, but also its superiority in health-giving and curative properties over all other kinds of milk. Apart from normal dairy and household usages, goats' milk is invaluable for babies, delicate people and invalids who cannot digest other milk, healing and curing where even drugs and high skill fail.

The goat industry is a growing one, with many sides to it. In this little book it is not my aim to deal with pedigree breeding or exhibiting, or commercial dairy goat-keeping, but rather to consider goat-keeping from the view of the ordinary family where a goat or two are required to provide sufficient milk for the household, and perhaps enable a little butter and cheese to be made in addition.

Family goat-keeping, in fact, with the goat as much a pet and part of the family (which it will surely become!) as it is the family's dairy shop, providing first-class food and drink, health and enjoyment, at absurdly low cost.

That this is possible in the case of an ordinary family, knowing nothing of practical goat-keeping at the outset,

1

has been proved by the writer and hundreds of others who have taken the plunge in this new venture.

Let me repeat, so that there will be no misunderstanding, that I am dealing with family goat-keeping. Not specifically a smallholder's or cottager's family with ample pasturage, woods or commons where the goats can roam, but the ordinary town or suburban family who have a normally good garden or allotment, and who may or may not have facilities for taking their goats for walks along hedgerows.

The idea that rolling acres are needed for goat-keeping has been quite disproved. Just one of many instances known to the writer is that of a goat-keeper who has kept goats for 14 years in a space of approximately five yards square—healthily, contentedly, and profitably (for a goat's contentment is reflected in the milk pail) and who from one goat alone has been getting up to $1\frac{1}{2}$ gallons of milk daily. There is further photographic proof concerning the possibilities of back-garden goat-keeping facing page 3.

To be able to provide free range may be an asset but is not by any means essential; and the same applies to the tethering system of keeping goats, which, thoughtlessly adopted, can bring so much suffering and loss. It is no more ridiculous to talk of back-garden goat-keeping than of back-garden poultry keeping. The goat is renowned for cleanliness almost amounting to fastidiousness. " Smell " is no more existent with a well-cared-for dairy goat than it is with a well-managed dog.

Family goat-keeping, therefore, opens up a vast new field of enterprise and, may I say, national service ? It is now everyone's duty to make his family as self-supporting as possible. When this can be done with the most essential of all foodstuffs, milk, every one who can possibly do so should see to it that he has a " goat in the garden," and further help Grand Old England on to victory.

This British Alpine
kept entirely in
small back garden
has been yielding
gallon of milk
day.

A Real Family Goat.

A Simple Garden Goat-house.

Once the garden
tool shed; now a
goat stable — and
very comfortable,
too, says Nanny!

CHAPTER II.

Getting Your Goat.

FIRST you will wish to know what type of goat to get, where to get it, how much to pay for it, and so forth.

Before the war, goat-keeping had reached a degree when the bigger breeders considered that a goat giving anything less than five pints as her average was not worth keeping. Even so, a great many people with shorter purses found a two- or three-pint milker a good investment. And she is even more so in wartime.

It has been said that a small milker is as costly to keep as a good milker. This is not true. Normally, a two- or three-pinter will produce on pretty well what she can pick up for herself in summer. The production of yields higher than this entails the feeding of concentrates and this obviously puts up costs of production.

To economise on concentrates must now be every goat-keeper's aim, natural foods being used as much as possible. The wealthiest breeders even have had to modify their ideas in this respect.

Very true is it, however, that a scrub will not yield so well on the same rations as a well-bred goat of good milk strain. It is true, too, that a poor goat will cost *more* to keep than a good one, if by the former is meant poor in health and condition and upbringing.

My advice, then, is to get the best goat you can afford, and if you can afford only a two- or three-pinter, then, providing she is perfectly healthy and hardy, don't despise her, for as a result of a season or two's breeding you can produce from *her* even better and still better milkers.

A friend of mine started with a couple of scrub goats he bought for £1. He obtained an average of three pints between them. Feeding cost him roughly a shilling a week. He was well content with the saving in his milk bill—21 pints of milk weekly for 1/- which, bought at the door, would have cost him at least 5/-.

3

One of the goats I once bought myself was a half-pedigree, the daughter of a scrub female mated to a fully pedigree stud male. She cost £3/10/0. She kidded in April, and from then until September gave me an average of seven pints daily, then gradually declined through autumn and winter to not less than five pints. Her feeding cost me 3/- a week. The returns were, apart from some useful kids, and a daily output of splendid manure for the garden, 49 pints of milk a week—12/- worth if bought at the door at 3d. a pint. I, too, was well satisfied.

Another friend of mine paid ten guineas for a pedigree goat. He was unfortunate. Her average was no more than my non-pedigree's. She cost about a shilling a week more to feed as she had been brought up rather' fastidiously. But wise mating, and the sale of good kids, soon put him on the right side.

The point is that there are good and bad milkers in both pedigree and non-pedigree goats. Pedigree or breed alone is insufficient recommendation for purchase.

A resumé of the various breeds will be helpful to those who know absolutely nothing about modern goats:

The Anglo-Nubian goat is the Jersey cow of the goat world, giving the highest butter-fat (cream) percentage of all goats, if not such a high yield as the Swiss types. A big animal, with a fine skin and glossy coat of varied colour, its very distinctive features are a Roman or camel-like nose and pendulous ears. All other breeds have prick ears. This breed is generally considered unsuitable for cold climates. While most affectionate and intelligent animals, they are noisy, which is a point to consider when thinking in terms of back gardens.

The British Alpine is a breed evolved by English breeders, a big animal, black with white or fawn markings, smart of appearance. A goat well known for its placid, quiet nature, and a heavy yielder of milk with averagely good butter-fats. Needs rather more exercise than smaller goats.

The Saanen is the white Swiss breed; a small, affectionate, heavy milking and long-lactation animal from which has been evolved the British Saanen. The latter, also pure white, is a larger, more heavily-built goat, a very heavy milker with excellent length of lactation. This breed has broduced two world goats' milk yield record holders.

The Toggenburg is a small, gentle and quiet Swiss breed, a good milker consistent in yield and excelling in long lactation. From it has been bred the British Toggenburg, which is larger and finer coated than the Swiss, with colour varying from the drab and white of the former to dark chocolate and white. The British type also has a slightly raised bridge to the nose; the Swiss type a straight or dished facial line.

Then there is what is known as the British breed. Usually of Swiss type, these are goats in which breeding, or parentage, colour, or "points" do not quite qualify them for registration in other breed sections. Excellent milkers and attractive animals, they are highly esteemed.

In all the above, the animals may be horned or hornless. Few pedigree horned goats are to be seen nowadays, however.

There are two other breeds indigenous to this country— the English (or Old English) and the Welsh. The former is small, short-legged, thick-set, horned, and very hardy. It is a great pity that little attempt has been made to improve yield with this breed for, while they are not heavy milkers, they are steady milkers, especially in winter, and butter-fats are excellent. The original Welsh is short-legged, small-bodied, long-haired and long-horned. The Improved Welsh Goat is now much in evidence in the Principality, of better size and type, greater milk yield, and shorter hair, thanks to careful breeding.

Except for the Anglo-Nubian, it is immaterial whether a goat has or has not tassels, or fleshy appendages on the neck. They have no connection with milk yield.

There are cross-breds to be seen, of course, such as Nubian-Toggenburg, and half-pedigrees will be found answering closely to specified breeds. They are frequently advertised as, say, Saanen type, or Toggenburg type, or B.A. type.

At the present time the average householder's requirement will be for a moderate but steady yield of five pints or so daily for the lactation, from a healthy, hornless, attractive quiet animal capable of yielding on wartime rations. This average means that at the peak of lactation soon after kidding, the goat will give for some time seven to eight pints daily. Then is the time you will be rearing kids, making butter and cheese, and feeding surplus skim milk to your poultry, rabbits, or back-garden pigs.

A good half-pedigree goat will cost from £3 to £6 if purchased from a reliable source. A pedigree milker will cost considerably more, breeders roughly valuing their stock at £1 per lb. (nearly enough one pint) of milk yielded at full flush up to 10lbs.; 30s. per lb. for goats giving 10 to 15lb. daily. There is also reputation to pay for.

If a definite breed is desired, choose that, if possible, which is favoured in the district. It will be acclimatised, and there should be good males available for breeding.

Be careful where you buy. See the goat if you can, or purchase on approval only. Beware of the advertisement which offers gallon milkers from 30/-. Such do not exist. Choose a recognised breeder, or make sure you get a personal recommendation or unbiased outside opinion. You can obtain advice, and the addresses of genuine goat-keepers, quite freely.

Should you buy a milking goat, an in-kid goat, or a dry goat ? is a question you may raise. The in-kid goat is the best proposition. She has time to settle down with you before she comes into milk. Journeys and new quarters upset the yield of milkers; it may be difficult to bring back. A dry, unmated goat has nothing to compensate for her cost of keep, and a mate has to be found for her which may possibly incur trouble and expense. However, we cannot pick and choose in wartime, and as goats are so scarce many people will have to make the best of whatever is offered.

Initial expense can be saved by buying a kid or goatling. Here again the cost of rearing and keep before there are any returns must be considered. Fortunately, there are inexpensive and simple methods of rearing kids into good milkers.

A mated goatling will not be found the best "buy" for a beginner. As a first-kidder, she will have to be broken in to milk. Best for him to gain experience with a goat at her second or third kidding and already well accustomed to being milked.

A goat's best age is from four to six years, at her third or fourth kidding. Her yield then diminishes at each kidding, and usually drops to meagre proportions after nine years. The natural span of a goat's life is about 12 years. Several cases have been recorded, however, of goats living up to 18 years and yielding a pint or so daily.

It is now usual for well-bred goats to milk continuously for two or more years without an intervening kidding, the yield reducing in winter and picking up again in spring. Such goats are very useful for those who don't want the bother of dealing with kids every year.

One goat may be found sufficient for a time, but you will surely experience the desire for two or more later on. Make your plans accordingly. Arrange for one, if you wish, but plan for two. Don't, for instance, get a house which will only hold one. There is very little extra expense in a two-goat house.

Two goats will ensure a satisfactory milk supply all the year, mated at different seasons. Three goats will be more than adequate if the system is adopted of mating one in September to kid in February; another in January to kid in June; and the third left unmated to milk through, the order being reversed the following year.

In any case, goats are companionable to extreme, keeping happier and yielding better in company. A lone goat, especially if bought from a herd, will fret, or will attach its affections to its owner, sometimes to embarrassing extent, following him or her around like a faithful dog—which isn't always convenient !

Finally, it should be pointed out that, unless you are buying a youngster, the goat you obtain should be used to the conditions it will meet with you. An old goat accustomed to free range will not do well confined.

CHAPTER III.

Your Goat's House and Run.

IN the stalling or intensive system of goat-keeping what is required is sufficient space to enable a small exercise yard to surround or be attached to the goat house. The goat can be fed in its house in bad weather and in the " yard " when it is fine, racks or netting bags being suspended from the sides of the fence, to contain hay, greenfood, etc.

The goat has free entry to its run in all weathers, can obtain exercise when it likes, and can do no damage to precious garden plants and trees.

It is an economical system, for a goat is a notorious waster of food if it can pick and choose ; and even many large breeders have adopted it in preference to giving the goats free range, despite the fact that more than sufficient range is available.

Goats, being playful and adventuresome, will never suffer boredom if a few boxes and a plank or two, or a mound of earth, form part of the " yard's " equipment. Believe me, such a simple gymnasium will keep the goats delighted for hours—and you, too !

The run for the three goats of an acquaintance of mine is the space between the side of his garage and his fence. The house is a converted tool shed.

The size of the goat yard will depend upon individual circumstances and the number of goats. A space 5 yds. x 5 yds. will be ample run for several goats.

Naturally, if the goats *can* be taken for walks, so much the better. If they can be taken to waste building plots, roadside verges, commons, copses, and the like for feeding —so much the more saving for your pocket.

Taking the animals along hard roads, too, is an excellent way of keeping their hooves in trim.

The goat-house yard can be fenced with 4ft. high sheep netting, cleft pale fencing, or posts and rails. Ordinary wire netting is unsatisfactory, as the goats delight to rub against this, bulging and weakening it. The angle-iron

8

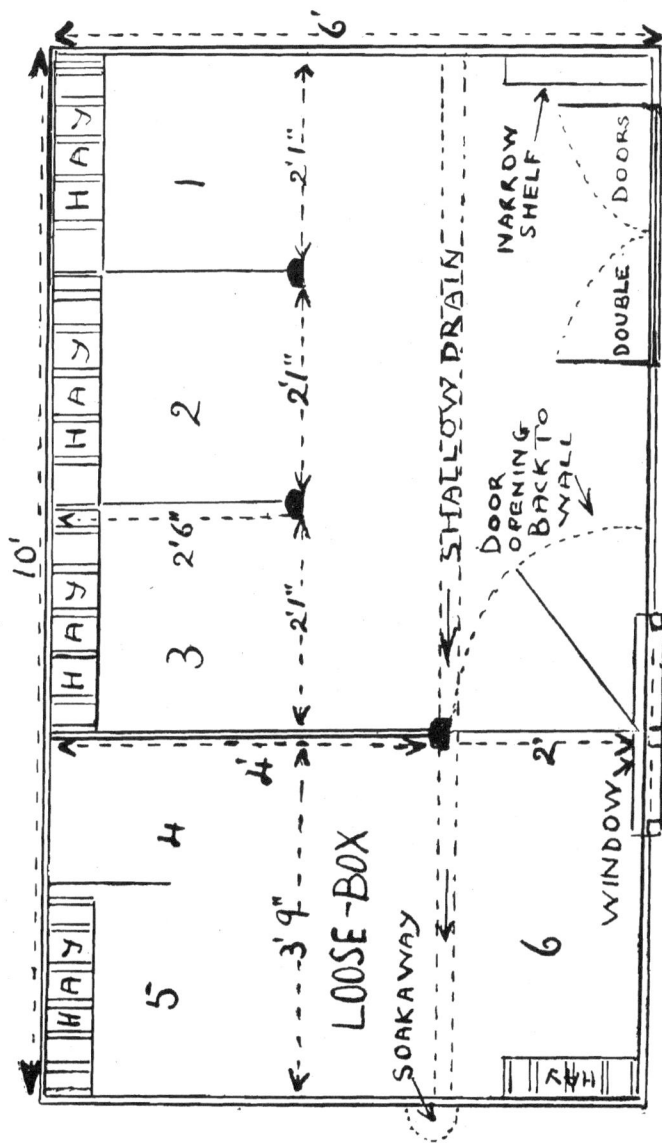

SCALE 1" = 2'

A Practical Family Goat House.

1, 2 and 3—Stalls for three milkers.

4, 5 and 6—Temporary "tie-ups" for growing kids. Small, low partition between 5 and 6 removable for kiddings, etc.

sides of bedsteads make first-class posts for cleft fencing
or sheep netting.

Coming to the house itself. I have seen goats living in
such luxury as I would not hesitate to partake of myself.
I have also seen goats whose only accommodation was to
bed down under the cart in the cart-shed. With our
family goat we must strike a happy medium.

Most beginners desire to know the simplest and least
expensive accommodation needed for one or two utility
goats, but it is a mistake to be too economical at the start,
the result frequently being greater expense later on. It
must not be overlooked that the goat is a strong animal,
and cheap, packing-case material will have a short life.
Walls, partitions, etc., must be sturdy.

The fullest use can, and should, be made of any existing buildings such as outhouses, stables, sheds, garages,
etc. They can be adapted readily to the purpose in view.
Where nothing of the type is available, a strong wooden
poultry-house or garden shed will make a good " stable "
if it is draught-proof, and if extra windows are fitted as
necessary.

In my own early days I used with success a converted
garden shed of lean-to type, bought for under £2 from
a sectional building dealer. With the door sawn in halves,
extra heavy felting on the roof, and one or two " fittings "
it made an admirable little house.

More goats can be kept in a small space if they are
stalled side by side, like cattle. Single stalls are usually
approximately 4ft. long and 2 to 2½ft. wide, divided with
partitions 4ft. high in front, sloping to 2¼ft. at the back.
The partitions are obviously to prevent the goats turning
round and soiling near the heads of the stalls, so that any
material that is available can be used for their construc-
tion. I have seen very good stalls made of old bedsteads. In another case I have seen stall sides which
fitted into slots, so that they could be taken down in a
moment either for cleaning out the shed or increasing
space.

The sketch gives details of a satisfactory type of stall
partition.

From my own experience I consider that the ideal type
of house for the small goat-keeper is one containing two
stalls and a loose-box. A loose-box is literally a section
of shed boxed off to a height of 4ft., in which the

goat does not have to be chained up. If a corner of the shed is used, a further two sides only are required, in one of which should be the entrance door. The loose-box will be found invaluable for " maternity " purposes—when one or other of your goats is kidding, for the accommodation of kids when the mother is back again in her stall and, when not in such use, for the handy storage of hay and sundries. A box should not be less than 4ft. by 4ft.,

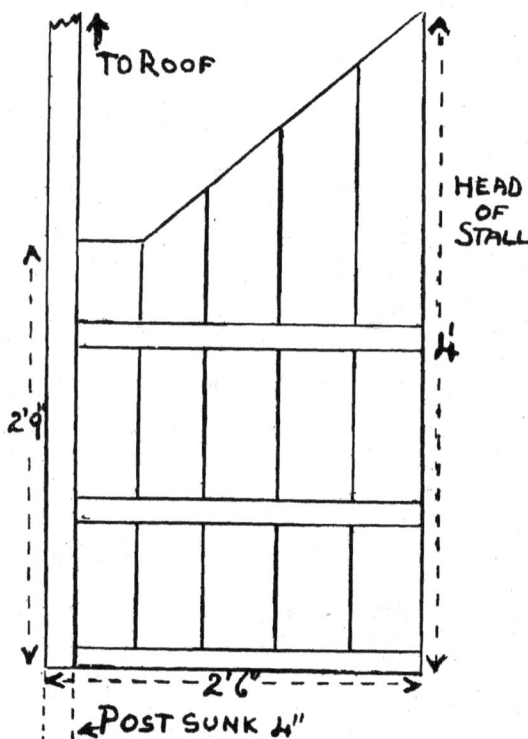

the minimum floor area required by a goat in such being considered as 16 sq. ft.

A plan of a small goat house as actually used under the stall-feeding or intensive system, faces page 8.

If the shed is large enough it pays to make the stalls double-width, so that the goats can be stalled in pairs. Thus, when you have made your two stalls you can in time accommodate four goats, as you feel the need to expand. Paired stalling not only saves woodwork, which is a great consideration these days, but also gives the goats that companionship and healthy rivalry in feeding to which the milk pail responds.

It is a convenience, where there are several goats and the required space, to have a passage round the head of the stalls along which the attendant can move for feeding without having to push his way past the goats; but with a small concern the finer points of goat-keeping can be left to more propitious times.

The goat-house should be about 6ft. to the eaves, or sufficiently high for the attendant's comfort, and possibly to provide a place overhead for the storage of dried branches, etc. Ventilation should be as high as possible to prevent draughts, and there should be foul air outlets of grids or louvres fitted at the highest point.

Light and sunshine are important for health. There should always be a window on the south side of the goat-house. If windows open from the top inwards, hopper fashion, there will be no direct draught on the animals.

Concrete flooring, sloping to a gully which itself slopes away to a sump outside the house, makes a sound job of drainage; and no bedding need be used for the goats, but each stall and loose-box should then have false floors of wooden battens nailed to runners on which the goats stand and lie without coming in contact with the urine and droppings which fall through, and preventing their delicate udders being chilled by the cold concrete. The false floors should be removable so that they can be taken outside for sweetening in the sun.

Wood floors in small sheds should be covered thickly with sawdust or peat moss litter (preferably the latter as it is better for the garden), to absorb urine, over which should be placed straw or waste hay litter.

The fittings required are quite simple. The hay rack is the most important. This can run the whole length of the head of stalls, 2½ft. from the floor, measuring from the bottom. One rack can be set between a pair of loose-boxes. Convenient dimensions for racks are 2ft. wide, 18ins. deep, and 12ins. across the top, tapering in to the bottom. The front should consist of rounded wood rails (projections might hurt the goat's face) set about 2in. apart, the sides should be enclosed, and if there is a hinged lid to the top the goats will be unable to stand up and pull the choicest hay from the top.

Stout cord feeding nets, specially made for the purpose, may be used, but their life will be shorter than a strong wooden rack. Some goat-keepers use racks with large mesh wire-netting fronts.

Concentrate rations and "small" food such as apples, carrots, cabbage leaves, etc., will be fed in ordinary gallon buckets brought into the house. Bucket holders which fall flat against the wall when not in use can be bought. Another idea given me by an experienced goat-keeper, to prevent buckets being tipped over and precious food wasted, is to obtain from an ironmonger large (3in.) spring hooks. Fasten with large staple into wall of goat-house about 2ft. off the ground. The feeding bucket handle can be very quickly pressed into the spring hook, and it is impossible for a goat to get it out. The staple should be in the corner so that the bucket fits in nicely.

Yet another suggestion is to use two-section orange boxes as receptacles for buckets.

A corner bracket may be fixed up to take a mineral salt-lick or lump of rock salt. Personally, I place this excellent aid to health on top of the concentrates in each goat's bucket.

Goats in stalls must be fastened by a short length of chain from the collar, spring-hooked on to a ring slipping up and down an 18in. iron bar or bolt secured to the wall with a couple of eye-bolts. This gives adequate freedom of movement.

Normally, the small goat-keeper does not find it advisable, or necessary, to keep a male goat, for reasons which will be found explained in the chapter on Breeding. But in abnormal circumstances such as these it may be very well worth time and trouble to keep a "billy."

He must be housed right away from the milkers—never on any account with them. However well kept he may be there is always a certain amount of odour with a billy in the breeding season. You do not want to risk him "tainting" the nannies. Moreover, whether you are building a house specially for him or adapting an existing shed, you should arrange the feeding of him so that he cannot rub round you, for you may not always want to go and put on an overall before going into his house.

The sketch on page 13 gives an idea for a small, very simple house for a single male. It may be seen that here he can be fed, watered, and given his hay and greenstuff from a feeding passage into which he cannot get. The front could be of substantial battens sufficiently far apart for him to get only his head through. Two pails on the

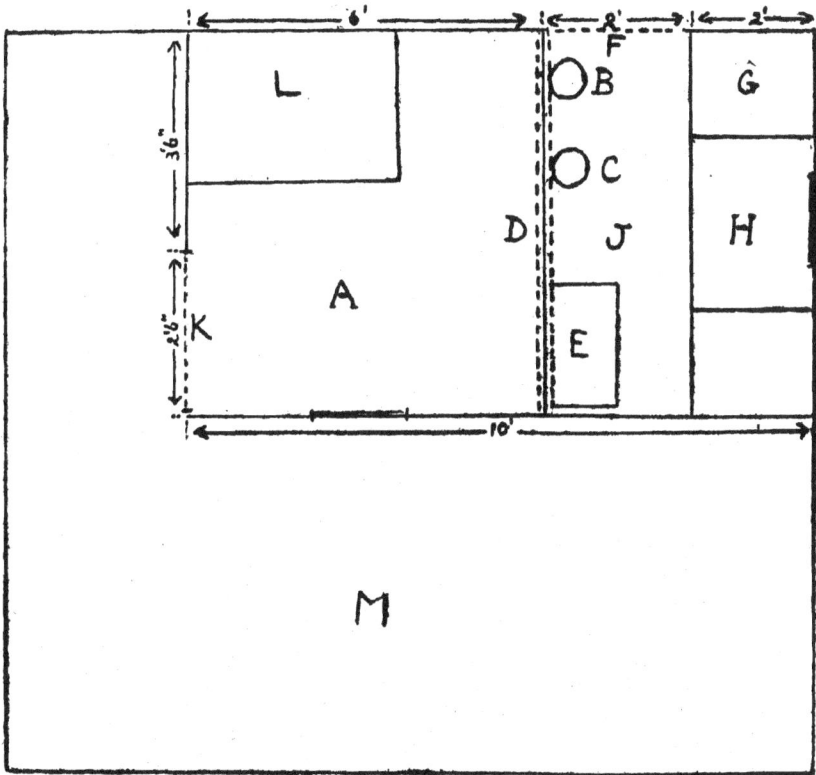

KEY TO DIAGRAM:

A—*Actual house.*
B—*Food pail.*
C—*Water pail.*
D—*Battens.*

E—*Hay box.*
F—*Door.*
G—*Corn bin.*
H—*Hay platform.*

J —*Feeding passage.*
K—*Billy's door.*
L—*Sleeping shelf.*
M—*Yard.*

floor outside this " barrier " would do for his food and water, and a large box fixed to the ground would serve for his hay and cabbage leaves, etc.

This is only a simple idea. A person handy with tools could give it many little refinements. At the back or side of the house must be a door for you to get into the billy when necessary, and for him to get into his run or yard which is an essential part of his house.

CHAPTER IV.

Feeding Your Goat.

You may consider your goat primarily as a wartime factory for the conversion of waste produce into milk. It is not *quite* as simple as that, though it is a captivating thought. It depends on the waste.

Contrary to comic paper conception, a healthy goat will not eat *anything*. A starved goat will; so will a starved human.

The goat's general cleanliness of habit extends to its feeding which further enhances its value as a dairy animal. Above all things, a goat desires clean food, and does not like other animals' or people's leavings.

If I give my goat a quartered apple she will eat it with relish. If I take a bite out of an apple and hand her the remainder, she will refuse it.

Perhaps I spoil my goats (though I think that pays), for I always carry a few toffees or biscuits in my pocket when I visit them, and they will nuzzle me and my pocket until I produce one. Should that toffee drop from the goat's mouth to the floor, however, she will not touch it again. I have tried disguising the toffee in a fresh paper, but even then it is refused. If hay, pulled from the rack, is dropped, it is never eaten.

A goat may eat and drink from the same buckets as her own particular stable companion, but never after any other animal.

These points are made to press home the type of animal with which you are dealing, and I repeat, they are definitely in the goat's favour. One can really fancy the milk from such a dainty creature.

There are scientific methods of feeding goats for high yields, and there are rules for balancing rations so that daily the goats receive their correct proportions of proteins, carbohydrates, vitamins, and the like. I do not propose to touch on this aspect of feeding, for such information can be obtained from many authoritative works. Besides, the war has complicated this subject, and many and strange

14

A Kodak Snapshot.

It is a popular fallacy that goats will eat anything. True, they are fond of tit-bits and have a sweet tooth, but in every respect they are dainty and fastidious over their food, and will never touch anything that is soiled. You can really fancy the milk from such a clean animal.

are the foods and mixtures now fed to goats which would have given rise to horror before the war. Yet the results are found to be extremely good !

Rationing of animal feeding stuffs is also in force now and places further restriction on what we can and cannot feed to our goats. Goat-owners are, however, much more fortunately placed than other small stock keepers—in respect of feeding stuffs goats have been allowed, by the Ministry of Food, the same priority as is enjoyed by dairy cows. In order to obtain feeding stuffs the new goat owner must register with his County War Agricultural Executive Committee, the address of which is the county town.

I shall tell you of the various corn foods (concentrates), green foods, roots, etc., you can feed to your family goat, how to estimate her requirements, and so on. But do not, I beg of you, follow me by rule of thumb. Study the individuality of each goat not only as to her food fads and fancies, but also as to her appetite. Try to feed as much home-grown food as possible. Also, do not be afraid to experiment with whatever foods are available in your district. Perhaps, for instance, you will be in a position to obtain brewery " screenings " which, soaked, can be added to mixtures in place of oats, with excellent results.

The Champion British Saanen which broke the world's officially recorded milk yield was fed on a simple con- centrate ration consisting of 3 parts flaked maize, 2 parts broad bran and 1 part crushed oats, by measure. Of course, in addition she had tit-bits such as baked bread, and hay and greenfood. On this ration she produced an average approximating to 1½ gallons of milk daily for 365 days.

And why I say don't be afraid to experiment is because I know of a cottager's scrub goat which, on such simple feeding as grass, cabbage leaves, and yellow meal boiled with wheatings in a gruel, has given 6 to 7 pints of milk daily.

As with dairy cattle, so with dairy goats, feeding is considered in two parts—maintenance ration and produc- tion ration. A normal goat of approximately 135-140 lbs. weight will maintain herself in health and condition and yield two or three pints of milk daily on what she can pick up for herself during the summer on good browsing or on a daily menu of 3 lbs. hay, 3 lbs. cabbage or roots

FOOD.	MILKERS, IN-KIDDERS, AND GOATLINGS. Parts by Weight (read each column down for complete ration).																MALES.	KIDS (from one month old).
Bean Meal	—	—	—	—	—	—	1	—	1	—	—	—	—	—	—	3	—	—
Beans, Kibbled	—	—	—	—	5	5	—	3	—	—	1	—	3	2	3	1	2	—
Beet Pulp (Dry)	—	2	1	2	—	3	—	—	1	2	2	1	2	—	5	—	—	—
Bran	1	—	1	1	—	—	1	—	—	—	—	2	1	—	—	—	—	1
Bread (stale)	—	—	—	—	—	—	—	—	—	—	—	—	—	—	—	—	—	—
Cotton Cake (decorticated)	—	—	1	—	—	—	—	—	—	—	—	—	—	2	—	—	—	—
Cottonseed Meal	—	—	—	—	—	—	—	—	—	—	—	—	—	—	—	—	—	—
Dairy Nuts	—	—	—	—	—	—	1	1	1	—	—	—	—	—	2	—	—	—
Dried Grains	—	—	—	—	—	—	—	—	—	—	—	—	3	—	—	3	—	—
Earthnut Cake (decorticated)	—	3	—	—	—	—	1	2	1	1	1	—	—	—	—	—	—	—
Lamb Food	1	—	—	1	4	7	2	2	1	1	1	4	3	—	—	—	—	2 } (crushed fine)
Linseed Cake	—	—	—	—	—	6	1	2	1	1	1	—	—	—	2	—	—	1 }
Maize (flaked)	1	3	2	1	—	—	1	3	1	1	1	—	—	—	—	—	—	—
Maize Gluten Feed	1	1	—	1	—	—	1	—	1	1	1	—	3	3	3	—	1	1
Maize Meal	—	—	2	1	—	—	1	—	1	1	1	4	3	—	—	—	—	—
Middlings	—	—	—	—	2	—	—	—	—	—	—	—	—	—	—	—	—	—
Oats (crushed)	—	—	—	—	4	—	—	—	—	—	—	—	—	—	—	—	1	1
Roots (pulped)	—	—	—	2	—	—	—	—	—	—	—	—	—	—	—	—	—	—
Soya Bean Meal	—	—	—	—	—	—	—	—	—	—	—	—	—	—	—	—	—	—
Soya Cake	—	—	—	—	—	—	—	—	—	—	—	—	—	—	—	—	—	—
Wheat (flaked)	—	2	—	—	—	—	—	—	—	—	—	—	—	—	—	—	—	—

(or their equivalent), and 1-1½ lbs. of concentrates (corn or meal foods).

To maintain a yield of milk over 2-3 pints, the goat must have an additional production ration of approximately 6 ozs. of concentrates per lb. (roughly a pint) of milk. Thus if your goat gave six pints of milk you would give her roughly 2½ lbs. of concentrates daily. But experience will teach you that the actual amount depends upon individual appetite and capacity for conversion of food into milk, and upon the kinds of other foods available.

The appended rations chart will show you how to make up complete menus for your goats, according to the foods you can obtain in your district.

The complete concentrate ration should be divided into two parts, one fed in the morning and the other in the evening, at strictly adhered to times. You will soon learn how many tinfuls or mugfuls of certain meals make a lb., so that there will not be the need to weigh out the meals separately each day. They should be fed dry as a rule.

Variety in greenfood feeding is one of the secrets of keeping goats well and contented. And there is plenty of variety to be obtained even in one's own garden or neighbourhood—cabbage, kale, prickly comfrey, chicory, tares, artichoke tops, pea haulm, sunflower foliage, beetroot and turnip tops, grass clippings from the lawn, sheep's parsley, cow mumble, docks, nettles and other common weeds, fruit and ornamental tree prunings, hedge prunings, rose bush prunings, gorse, heather, brambles, briars, nut, and so on.

Evergreen trees, hedges and shrubs should be avoided, however, as the majority are poisonous to goats. Three exceptions are the evergreen honeysuckle, *lonicera nitida,* a quick-growing hedging plant of which the goats are fond, holly, and ivy without the berries.

In winter, when greens apart from hardy kales are scarce, the goats can have such roots, sliced or pulped, as kohl rabi, carrots, mangolds, turnips, and cooked potatoes and artichokes.

Greens and roots should be fed at midday, and changes may be rung by pulping them and " drying them off " with bran or middlings, to eke out the concentrates.

Hay is an essential, and in my opinion should always be before the goats. There is nothing like good, sweet hay for producing milk. If you cannot make your own hay, it can be bought by the truss, costing from 2/6 to

c

4/6, and a truss will last a goat about a month. I have frequently brought a truss of hay home on the back of my old car from some outlying farm.

There is good and bad hay. The best *is* best—it will produce more milk, and less of it will be wasted. In order of preference, goats like sanfoin, lucerne, nettle, clover, and meadow hay.

Goats can receive their " minerals " via a block which they lick (and obtainable from any veterinary chemist's), or proprietary powder minerals which you mix with the food. You can also make up your own mineral mixture as follows: Sterilised bone flour, 5 lb.; finely-ground lime-stone, 5 lb.; common salt, 2½ lb.; and potassium iodide, 2 ozs. per 100 lb. of mixture—dissolved in a little water and sprayed over the mixture from a scent spray. 6 oz. of this mineral mixture should be used per 12½ lb. of concentrate mixture.

Goats must also have water. Milk contains 86.57 per cent. of water. Therefore plenty of water is the first essential in milk production.

The poor yields given by many goats is most frequently because either they do not receive sufficient water, or their " appetite " for water has not been developed. The latter is a point all kid rearers are advised to remember. All high yielding goats have a surprisingly large thirst.

Goats vary considerably in their reaction to water. Their fads and fancies must be pandered to, to induce them to drink more.

Goat keepers must remember that no goat will drink dirty water. Frequently she will not drink after another goat from the same bucket. Water that has been left standing in the goat house and is tainted with the ammonia from the litter will be refused.

I had a goat which would not drink cold water even on the hottest day of summer. But she would never refuse warm water at any time.

When goats change hands, change of water often upsets their desire for it. Used to soft water, they object to the new supply of hard water, or *vice versa*.

Water requirements need to be studied along with the diet. Thus a goat on a hay and concentrate diet requires and will drink more water than one receiving ample succulent green food.

Goats can, and should, be trained to a watering routine

from an early age. Water should be offered them at set times every day throughout the year, until " watering " becomes a fixed habit. It is of little use leaving water always before the animals.

Finally, when a goat refuses water, the best inducement is to make the water warm, and drop in a little salt or middlings.

THE ROUTINE OF STALL-FEEDING.

The following is an account, specially written for me by Miss N. M. B. Smales, of how a goat-owner manages her small herd on the semi-confined, or stall-feeding, system:

Much has been written on the different ways of keeping goats, but I hope this will be of help to amateurs who are wanting to start with goats but are afraid they have not sufficient space for them. I have always kept my goats on the " stall-fed " system and have so far managed to keep them fit and well. Never have I had more than a very small patch of ground for their run or exercise yard.

All this last summer and winter I have had three milkers and reared three pedigree kids, two from the age of three weeks and one who came to me when she was about six weeks. The goat house is small and comprises three stalls and a loose box. The hay rack extends along the length of the stalls, and there is a similar one, only built nearer the ground to enable the kids to reach it, in the loose box.

The goats lie on creosoted boards which are only slightly slatted, being made from broken up bulb boxes, one to each stall and two for the loose box to facilitate handling.

The house, 10ft. x 6ft., is standing at the back of an enclosed space about 20ft. x 20ft., which is the only space available for the exercise yard.

The following details of daily routine, making allowances for weather conditions, etc., will show you exactly how my stall-fed goats are managed.

Summer routine—

7.30 a.m. Each goat receives half her daily ration of concentrates according to her personal needs, and her milk yield.

Milking starts immediately after all are fed.

8.30 a.m. If the weather is fine, every animal is turned out into the run. Branches of leaves are hung up for

them or cut grass or other fodder is put into their racks.
Cleaning out is done when convenient.

12-12.30.   If it is very hot and the run is not shaded,
all are put into the shed, green food is put in the racks,
and the goats are left until it is cooler.   I believe they
are far better lying quietly in their stalls contentedly
chewing the cud and making milk, than trying to get out
of the way of flies, or panting in the heat.

Have you ever noticed that a herd of cows out at
pasture will nearly all be lying down during the middle
hours of the day?   It is while chewing the cud that the
milk is mostly formed.

Of course, if the weather is typically British and there
is no very hot sun, the goats can remain out in the run
all day, as long as the shed door is left open for them
to go in and out as they wish.

6 p.m.   Milking time, when each animal receives her
last ration of concentrates.   As soon as they have finished
this I usually take them all out for a walk in the lanes
and droves, where they can eat and browse to their hearts'
content in the cool of the day.   We are generally out
for 1 to 1½ hours.

Upon their return the goats are put to bed, and if I
do not think they have had enough, I " top up " with hay
or green food.

Winter routine—

8 a.m.   First feed of concentrates, and milking.

9 a.m.   Racks are filled with hay or ivy.

10 a.m., or as soon as it is warm enough, the goats
are turned out into the run.

Cleaning out takes place when convenient.

12-12.30.   The goats are put into the shed and given
cabbage leaves, roots, or vegetable parings, etc., when
available.   If not, a small feed of hay.   If the weather
is suitable they are turned out again about 2 p.m., but
if it is at all cold or dull they are kept in.

5.30 p.m.   The evening feed of concentrates, and
milking, followed by a full rack of hay.

At about 7.30 I go out and give them all a really warm,
mealy drink, which they appreciate very much.   This is
made by pouring boiling water on to a double handful

of weatings, or bran. Sometimes I add salt or black cattle treacle. Cold water is poured on just before I take it out to them.

In the winter, daily exercise is taken in the afternoons when possible—a brisk walk on the roads. At the time of writing the Foot-and-Mouth restrictions are in force here, and I have been unable to take my herd on to the roads for the past month. So the only exercise they have is a " scampede " once a day in a small patch of ground adjoining the house. The goats are remarkably well, and always seem happy and contented nevertheless.

Of course, this is only a broad outline of the management of stall-fed goats, and must not be taken as the only way. For instance, you may say: " It is all very fine, but I have to be away at work all day and can't give them a mid-day meal; nor can I put them in or let them out during the day."

All right! Make your own plan, by using your commonsense, and leaving the goats sufficient food in their racks to last them till you come home in the evening. As long as they are able to get into shelter from the heat, rain, and cold winds they will not hurt.

But a word of warning. Do make quite sure that the boss of the herd (and there is certain to be one) does not stand in the doorway and keep everyone else out! And again, it is always wiser to have an entirely hornless herd or an all horned one, for the ones with horns have a distinct advantage over the hornless ones, who may be kept away from food and shelter.

Then there is the fodder question—the branches I spoke about. If you have no time to go out and cut branches for yourself, it is worth giving a reliable boy or girl a few pence pocket money to get it for you, as long as he or she knows what is suitable.

Or again, if you have a fair-sized garden and can allocate a portion of it to be the " goats' garden " you can grow all manner of things for them. To mention a few: Carrots, mangolds, swedes for root crops, and all kinds of cabbages and kales; artichokes (both tops and roots), green maize, oats to cut green, besides all the things you normally grow for your own use—the outside leaves of broccoli, curly kale, savoys, etc. A patch or two of lucerne is invaluable.

Personally, I seldom go out for a walk without taking

a piece of rope, a clean, small sack, and a strong pair of secateurs. There are endless things you can pick in the hedgerows and at the side of the road, which can go into the sack, to say nothing about the apples left to rot, or the mangolds, swedes and turnips that have fallen from a badly loaded putt.

The secateurs will cut those lovely pieces of nut, ash and elm that are in the hedge, and these can be neatly tied into a bundle with the rope. Do use the secateurs whenever possible, and do not cut the hedges indiscriminately or you will get into the farmer's bad books for damaging his hedges.

On more than one occasion last year I cut quite small branches from a huge ash hedge, taking pieces here and there so that no one could say I was doing any damage. Imagine my disgust when some week or two later I again visited that same hedge, only to find that the hedger had been busy, had cut the whole thing down, and was gaily burning all the now dried up ash branches. I rescued what I could and came home wishing to goodness I had not been so conscientious.

So if you *can* find out what hedges a farmer intends to cut that year, he will probably be willing to let you have first cut.

If you have to leave your animals for any length of time during the day, do be careful there are no weak spots in the fencing, for they are almost sure to find it out while you are away.

Also see that there are no traps that a kid or goat can get her leg or head caught in.

If you are at all doubtful of these points, then I say emphatically: Leave the goats in their shed where you know they are safe. It is horrible to be some distance from home and the thought suddenly come to you that all may not be well at home.

To keep the kids amused an upturned box gives no end of pleasure as long as it is firm and will not tip over. Logs for them to bark are a great source of delight to young and old, and are good for them, too.

To sum up. Although it may appear at first sight to be a lot of trouble and that you are constantly doing odd jobs with the goats, under the above system a methodical person can so arrange the work that there is ample time for the usual household tasks, gardening, etc., and the goats' feeding times can be arranged to suit each individual household.

On the stall-fed system you know exactly what quantity of food each goat receives daily, and you are able to give that little bit "extra" to a particular animal which may need it, for no two goats can be treated exactly alike and careful observation soon tells you when each member of the herd is "doing" to the best of her ability.

Goats are always amenable and easily handled when stall-fed because they are as one of the household and treated accordingly.

One last point, when buying a goat which you intend to keep like this, it is advisable to buy one that is not used to a large field or "free range," because it may take a long time for her to settle down to the new way of living, and so cause disappointment to her new owner.

Experience was most costly upon one occasion—a goat was bought which had been running free wherever she willed, and, having horns, and being most self-willed, in no time had she charged clean through every wire netting fence she came to! Moreover, unused to high feeding, she could never digest corn feed and nearly "died on me" several times!

CHAPTER V.

Your Guide to Breeding.

WHILE I hope that you will start your goat-keeping with an in-kid nanny, so that questions concerning breeding will not trouble you during your first few months in your new venture, it is best, I suppose, to take first things first, and deal in this chapter with the mating of goats.

As far as milk production only is concerned, mating a nanny with any handy billy will produce the desired result. But where it is also desired to rear kids and improve one's strain, rather more is involved. Only pedigree males from proved milking stock should be used.

Goat-keepers have been particularly fortunate in the matter of breeding in that, until the war, the Government subsidised a Stud Goat Scheme by which cottagers, small-holders, and artisan goat-owners could obtain the services of first-class pedigree stud males at fees not in excess of 4/——fees which in the ordinary way might be two or three guineas.

While it is unfortunate that, owing to the war, the Scheme has had to be suspended, the majority of stud goat owners are still allowing services on pretty well the same terms; and such stud goats are to be found in practically every district. You can start with the veriest scrub nanny and by breeding consistently to pedigree males can have gallon milkers of your own within a very few seasons.

Normally goats breed only during the season extending from the beginning of September to the end of February. The period the kids are carried is approximately five months or 21 weeks. When you buy an in-kid goat, then, find out exactly when she was mated, and you can tell within a few days when the kids will be born. The following table will help you in this respect.

I said previously that goats *normally* breed from September to February inclusive. Goats *can* be mated before

24

BREEDING TIME-TABLE.

Breeding Season taken as from September 1st to February 28th. Period of gestation as 21 weeks. Allowance for leap year is made. Duration of Oestrum—2-3 days. Abortion most liable to occur at 5th, 9th, or 13th week.

Service Date	September 1 2 3 4 5 6 7 8 9 10 11 12 13 14 15 16 17 18 19 20 21 22 23 24 25 26 27 28 29 30
Kidding Date	January 26 27 28 29 30 31 Feb. 1 2 3 4 5 6 7 8 9 10 11 12 13 14 15 16 17 18 19 20 21 22 23 24
Service Date	October 1 2 3 4 5 6 7 8 9 10 11 12 13 14 15 16 17 18 19 20 21 22 23 24 25 26 27 28 29 30 31
Kidding Date	February 25 26 27 28 29 Mar. 1 2 3 4 5 6 7 8 9 10 11 12 13 14 15 16 17 18 19 20 21 22 23 24 25 26
Service Date	November 1 2 3 4 5 6 7 8 9 10 11 12 13 14 15 16 17 18 19 20 21 22 23 24 25 26 27 28 29 30
Kidding Date	March 27 28 29 30 31 April 1 2 3 4 5 6 7 8 9 10 11 12 13 14 15 16 17 18 19 20 21 22 23 24 25
Service Date	December 1 2 3 4 5 6 7 8 9 10 11 12 13 14 15 16 17 18 19 20 21 22 23 24 25 26 27 28 29 30 31
Kidding Date	April 26 27 28 29 30 May 1 2 3 4 5 6 7 8 9 10 11 12 13 14 15 16 17 18 19 20 21 22 23 24 25 26
Service Date	January 1 2 3 4 5 6 7 8 9 10 11 12 13 14 15 16 17 18 19 20 21 22 23 24 25 26 27 28 29 30 31
Kidding Date	May 27 28 29 30 31 June 1 2 3 4 5 6 7 8 9 10 11 12 13 14 15 16 17 18 19 20 21 22 23 24 25 26
Service Date	February 1 2 3 4 5 6 7 8 9 10 11 12 13 14 15 16 17 18 19 20 21 22 23 24 25 26 27 28
Kidding Date	June 27 28 29 30 July 1 2 3 4 5 6 7 8 9 10 11 12 13 14 15 16 17 18 19 20 21 22 23 24 25

and after this period, but " in season " signs are not so
prominent, and the results are not so certain.   Also, no
two goats are alike.   Some, coming early " in season,"
kid down in early spring (all to the good, the kids having
the best " growing " weather before them); but others do
not appear in any hurry, and in seeking their mates late in
the season kid in late summer or early autumn—these are
termed " winter milkers."

How to tell when a goat is " in season "?   Here, again,
no two goats are alike.   Some, very quiet, and possibly
shy ones, will show they are in season by a rapid shaking
of their tail, a restlessness of manner, and sometimes a
refusal of food.   Those that are giving milk will drop
their yield suddenly—a goat giving usually two or three
pints will one morning give about half a pint!   Of
course, the yield returns after the " in season " period
which may last from two to three days.

The noisy goats will bleat continuously, also shaking
their tails, although it may be a full 24 hours before
they are properly " in."

I think the best way to be sure is to walk quietly round
and try to get a rear view of the " suspected one."   If
under the tail there is a swelling, red and sore looking,
or bluish in colour, and a whitish discharge, or even trans-
parent, the goat should be taken at once to the male, if
you wish her mated.

Just a word of warning here to those who want to breed
good goats.   If in your locality, even as far as two miles
distant, there should be an undesirable little scrub, do not
leave your " in season " nanny loose or tethered.   Either
she will visit the undesirable one, or he, with that most
wonderful, unerring animal instinct, will visit her.   So
watch carefully every morning until the goat is mated
that no such accidents happen.

And when the goat has been to her chosen mate, still
watch.   She may " turn "—that is, come " in season "
again a week after mating, three weeks (the usual time
elapsing between each " season " until mated), at four, or
even six weeks.   Therefore a goat cannot be guaranteed
" in-kid " until she has been six weeks mated without any
further sign of " season."

It sounds a lot of bother, but it is not.   Any good
stock-keeper will have a " watchful eye " always on the
stock—and goats are valuable stock.

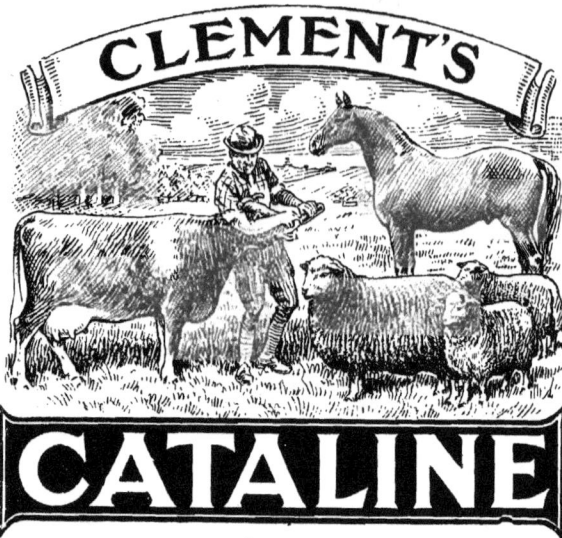

CHAPTER VI.

Kidding Without Worry.

To the beginner, I know well, the most to be feared part of goat-keeping is the first kidding experience. Anxiety increases at the approach of this time, and there is the fear that " everything will not be all right."

Such fears are absolutely groundless, as I and others have proved.

The following has been written for me by a breeder of very considerable experience, and I can vouch for every word of it: —

" When I started goat-keeping I knew very little indeed about goats. I started with an old non-pedigree nanny due to kid about a month after her arrival. I read all the literature on goats' kidding which I could get hold of, and the result was that as Pat's kidding approached I was simply terrified of what was going to happen, for the articles had told of all the complications which *might* occur so thoroughly that it seemed to me then that an easy, uncomplicated kidding was quite the exception.

" Pat kidded with no trouble at all, and in fact so easily that I was absolutely surprised. I have had many goats since old Pat's day, and have witnessed many, many kiddings, but I have only ever lost two goats through kidding, and both of these were old goats. Complicated kiddings I have found are decidedly the exception and I do not intend to deal with them here. The average healthy goat, not too highly bred, kids easily and is no bother. So, now feeling maybe, a little less worried over the coming event, what *can* you expect?

" The gestation period of goats seems to be about 150 days, but watch her carefully after the 147th day, for after then you may expect the kids at any time. Put her in a loose box, *untied*, with plenty of clean straw bedding. Don't leave food or water pails in the box in case of accidents. Watch her udder; if it is very large and tight don't be afraid to milk her before she kids, it will do no harm and may do a lot of good, but only loosen it; don't, of course, strip her.

"Towards the end of the gestation period the goat will look less bulky, her tail will seem to be carried higher, and her flanks will look hollow, and there will be a deep depression either side of her tail—in fact, the kids have dropped into the position from which they will be born.

"The first symptom is a thick-looking white discharge. The time elapsing between the start of this discharge and the actual birth of the kids varies considerably from an hour or two to several days, and so a watch must be kept carefully for further symptoms.

"The goat next becomes restless and will call out to you in quite a different tone from her usual bleat, a rather scared little noise in fact.

"The next stage is the change in the nature of the discharge. It will become a yellowish opaque one, rather like the white of an egg to look at. The kids are not far away now. By this time the goat will be scratching her bedding into heaps, lying down, getting up, remaking her bed and generally exhibiting every sign of uneasiness. She will generally go on like this for some little time until she lies down and starts to strain, slightly at first, but as her pains increase, more and more vigorously until the kid is born.

"As soon as the kid is born clean away the slime from its nostrils and with a finger wipe out the mucus from its mouth. The mother will quickly start to clean the rest of the kid and it will soon be on its feet looking for its first feed.

"If there are more kids to follow the goat will often start straining again even before she has finished cleaning the first one. Rub the kid down with a soft towel, then put him out of the way in a box of hay but where the mother can see it. The second, or second and third kids, as the case may be, will arrive quickly and with very little trouble. Clean their noses and mouths like the first one. All that remains now is the expulsion of the afterbirth. Generally this will follow quite quickly, but it is strongly advisable always to stay with the goat until you are *sure* the afterbirth has come completely away.

"The mother will be very ready now for a nice warm

bran mash and a drink of warm oatmeal with a dessert-spoonful of black treacle in it. The kids will be looking for their first feed; sometimes they are not clever enough to find the teats and just one lesson is needed. If the mother's udder is not sufficiently loosened by the kids, take a little out, but do not strip her for the first day or two until her full flow of milk has come in.

"Mother and family will now be ready to be left alone for a good rest and sleep, and so your worries are over, and you will face your goat's next kidding, I hope, with less anxiety."

CHAPTER VII.

The Kids.

WHEN your kids arrive (never forgetting that twins and triplets are quite common) make an immediate examination of them to determine if they are male or female. If male, and unless fully pedigree on both parents' sides, do *not* keep them with the intention of breeding from them later. Either kill them at once, or keep them for fattening for table.

Immediately after birth they can be killed, like rabbits, by a sharp blow at the back of the head. Or they can be put to sleep at an Animal Centre.

But criminal offence as it now is to waste food, *don't* be squeamish about fattening kids for the table. Kid meat is a real delicacy, and the kids can be reared cheaply on milk substitute. More about goat meat will be found in another chapter.

You will keep nanny kids, of course, and you now have confronting you the " problem " of rearing them. In the old days they were just left with the mother, and you took what milk you could when the kids had finished.

Now we are wiser and produce better goats by rearing the kids by hand, and we save the milk for ourselves by switching the kids over after a time to a good milk substitute providing all growth essentials, yet costing very much less than the value of milk to us.

Hand-reared kids become real pets; you can train them to your own particular way of life, and they grow up into well-mannered, quiet adults who will answer your every beck and call.

It is certainly more troublesome to rear a kid by hand than to let it run with its dam—but here is where the family comes in. Your children will clamour to feed the kids, and my own little girl has proved that the young are perfectly capable of feeding the young.

Here are the essential facts on hand kid-rearing. Naturally, they may need modifying according to individual circumstances: —

It is important, right from the first, not to allow the kid

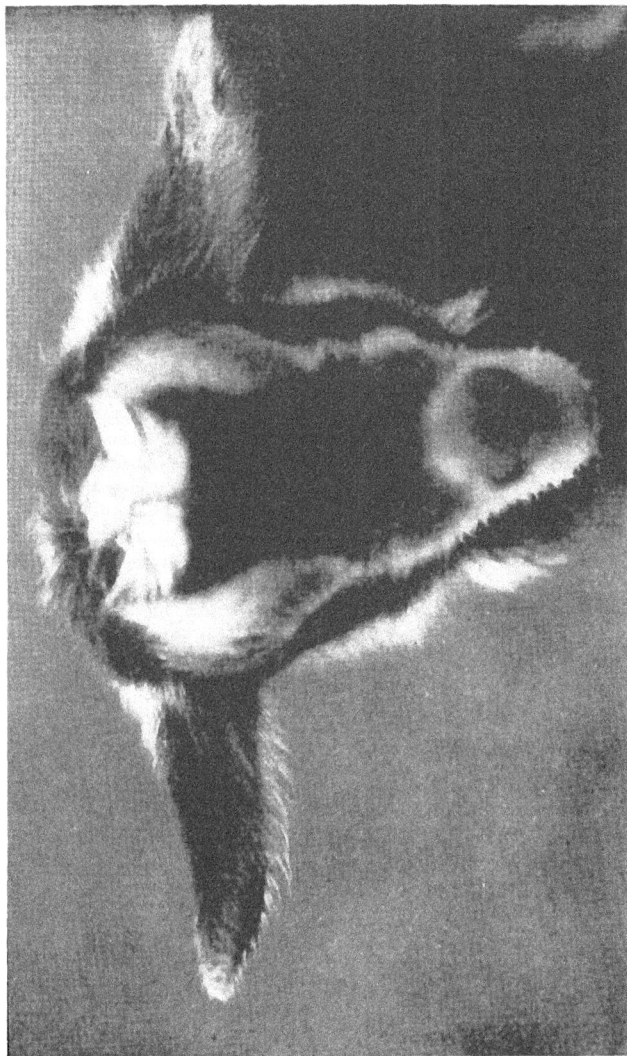

Full of charm, intelligence and high spirits, which they never lose as they grow up, who can resist the appeal of the kids to be loved and petted? The family goat-keeper has many happy moments in store when the kids come.

to suckle its mother; otherwise there will be great diffi-
culty in parting them later, or in getting the kid to take
to a bottle. In some cases rearers have found it practicable
to allow the kid to remain with its mother for the first
two days, and then start it on its bottle-fed career.

The milk which first comes from the mother is thick,
yellowish colostrum—specially designed by nature for the
first few days of the kid's life. The kid must receive this,
but if for any reason it is not available, a half-teaspoonful
of liquid medicinal paraffin should be shaken up in the
milk or milk-substitute bottle once a day. The mother's
milk changes in character after two or three days, and
by the fourth day it is quite normal for consumption by
human beings.

For the first few days you can feed from an ordinary
baby's feeding bottle with teat and valve, 8oz. size. At
four days you will find a tomato sauce bottle, holding just
about ¾ pint, ideal. The neck should be straight, with
no ledges where milk can collect to go sour. At about
six weeks onwards a quart size vinegar or lemonade bottle
with a straight neck is excellent.

Following the baby bottle, use soft rubber lamb teats,
enlarging the hole with a hot steel knitting needle.

The milk must be fed at blood heat, *never* cold. It
can be heated in two ways: In a saucepan over the fire,
or by putting the milk into a bottle and placing it in a
jug of hot water until it is the right temperature. Test
the heat of the milk on the back of your hand.

Generally speaking, a strong, healthy kid will take about
8ozs. of milk each feed till about ten days or a fortnight
old, increasing to 12 ozs. Kids should never be forced
to take their bottles. If they refuse a bottle, or not quite
finish one, no anxiety need be felt unless it is a constant
occurrence, or the stomach seems to be upset. The
quantity can be gradually increased until the kid is having
a pint per feed four times a day at three months old.

From birth to ten days or a fortnight the kid should
be fed five times a day, the hours between each feed being
even, except the last feed at night, which should be as
late as possible. At a fortnight one feed can be dropped,
but a little more milk given at the other feeds to make
up. Good times for the first fortnight are 8 and 11 a.m.,
2, 5, and 9 p.m. After this, 8 a.m., 12 noon, 4 and 9
p.m. are best.

Regularity is important, and once your times are fixed
they should be adhered to.   To allow a kid to become
too hungry, when it will take the bottle too fast, is to
invite tummy trouble.

Kids will nibble grass, greens, hay, and earth (which
is very good for them) from an early age.   They usually
begin to eat concentrated food at about a month old.   A
little bran, flaked maize, crushed oats, and the dust of
linseed cake should be put in a shallow vessel.   They will
soon learn to eat it.   As soon as they begin to eat this
solid food they will chew the cud.

A bunch of sweet hay should be put in a low rack or
tied up in a bundle and hung at a convenient height, when
the kids reach a month old.   Make sure they cannot get
their heads or legs tied up in the string.   Branches of nut,
elm, ash, etc., can also be hung up for them.   A very small
quantity of cabbage, kale, etc., may be given, but only
when quite dry and not touched by frost.

An iodised mineral salt lick should be available from
the start.   You will notice the kids going to their " salt
cellar " and having a good lick after a bottle.

Apart from bottle-rearing, kids can, of course, also be
taught to feed from a bowl or bucket, thus saving time
and expense with bottles and teats.   A drawback with this
method is that they may drink too quickly, become " pot-
bellied " and scour.   To induce the kid to drink, let it
suckle a finger dipped in the milk, and gradually draw its
mouth down to the milk.   A very few lessons like this and
the kid will drink of its own accord.

Kids must never be coddled, and must have plenty of
exercise.   In spring and summer they can stay out in a
small wired enclosure containing a box or barrel into
which they can creep to sleep.   Or you can let them have
the run of a shed or an ordinary chicken house, for shelter
and sleep, a large hay-lined box being inside for sleeping.
Any house used for kids must be draught-proof and leak-
proof, but be airy and well-lighted, and facing south,
sunshine being essential to their growth.

For exercise, a strong box turned upside-down in their
pen for them to jump on makes an ideal " castle," and
when they are a few weeks old a bridge made from two
upturned boxes and a plank rigidly fixed gives these
delightful little animals much pleasure.

A word about milk substitutes.   These can be used

when there is a scarcity of natural milk, or if you have a kid to rear, with no nanny available. There are several well-known makes on the market, and directions are supplied for their use. The cost works out roughly at 3d. per gallon. Normally, the substitute should be gradually introduced to supplement the milk, after the age of one month.

REARING A MALE KID.

If you have a well-bred male kid to rear, for the first three months of his life he can be kept with the female kids and treated exactly as they are.

Do not stint him with milk, but at the same time don't overdo it. Four pints of milk a day are sufficient for any kid, and three pints will be found to be enough for most kids. Too much milk weakens the digestion and the kid is apt to be " pot-bellied " and the flesh soft and flabby.

Encourage him to eat concentrates and hay as early in life as possible, give him as much concentrates as he will eagerly clear up (within reason, for some kids *will* eat more than is good for them, but I find as much as they eagerly clear up a good general rule) and as much hay as he can eat.

Exercise is most important for good growth but anyway while he is with the other kids he will get plenty if they are provided with boxes or a fixed barrel or something similar on which to play " king of the castle."

Teach him to lead well on a halter while he is young, to know his name and to come when he is called. These apparently minor points will save a lot of trouble when he is older.

It is a mistake to play with any male; jumping up and so on may be amusing when he is small, but remember that it will be you, and not he, who will know the difference when he weighs about one and a-half hundredweights !

The male kid will probably be used from six months onwards, and so during his first " season " he wants perhaps even a little extra care so as to keep him growing well. As the season approaches, his appetite will fall off badly; try and tempt him with a change of rations, or add a little treacle to his concentrates. Try bread crusts, flaked maize on its own, a little spice in his food, or anything you think he will eat. If, however, he definitely will eat

D

but very little, don't worry unduly, for they soon pick up again once the season is over. The great thing is to have your male in really first-class condition at the beginning of the season. See that a mineral lick is always available.

Don't treat a male as a complete outcast. If there are no goats in season he can quite well accompany the females on their walks with you, if he is equipped with a halter. Try to amuse him a little while he is alone in his yard, for boredom is the chief cause, I believe, of objectionable habits. Give him leaves to pick over, branches to bark, and a section of an old tree trunk—or something as strong —will be welcome to jump on.

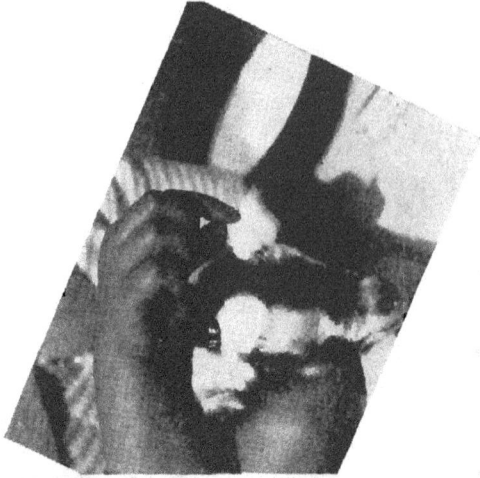

horns can be prevented by a simple "operation." Instructions are given on page 35, and should be studied in conjunction with these photographs.

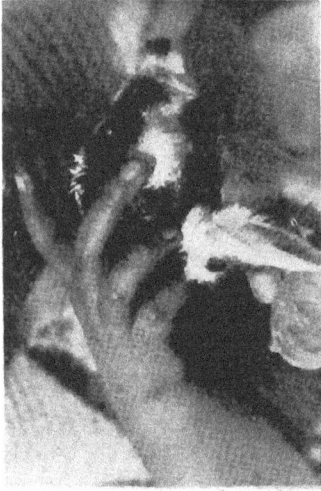

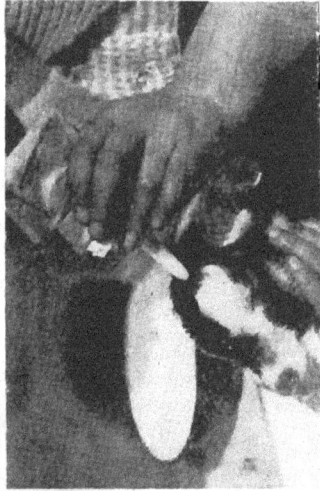

DISBUDDING

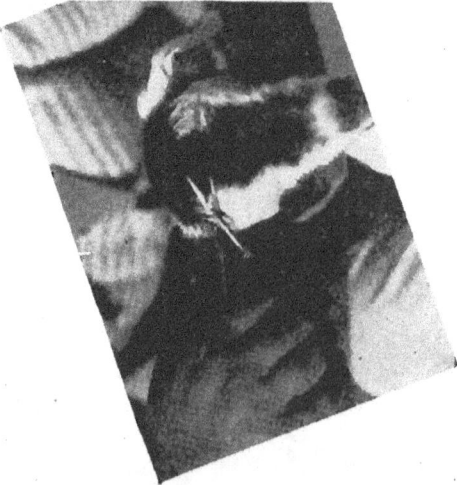

The family goat should preferably be hornless, for safety's sake. The majority of goats are now bred without horns, but if a kid is going to be horned, the growth of the

CHAPTER VIII.

Disbudding Kids.

WHETHER a goat has horns or not is immaterial from a milk production point of view. A horned goat is likely to be as good a milker as a hornless one. But it is a matter of practical advantage to have goats without horns, especially family goats. Accidents can happen even in play with the quietest animals.

The horns of goats can be prevented from growing by disbudding kids at a few days old. This operation can be performed by anyone who is prepared to take a due amount of care. It should be done at not later than three or four days old.

Your kids may not necessarily be going to grow horns. One or more of them might be, however, and you should be prepared.

A method of telling if a kid will grow horns is to wet the animal's head and smooth back the hair from above the eyes to the top of the head. A hornless kid's hair will remain smooth; otherwise, the " curls " which will be present will reappear. Head shape is also a guide. A horned kid will have a flat forehead, not rounded. Another method to determine if disbudding is necessary is to try to move the skin back and forth over the forehead. If quite freely movable no horns are developing; if the skin seems grown fast to a prominence horns can be suspected.

In time, of course, the " buds " can be felt developing, and later, little horns.

To disbud, all hair should be clipped away from the horn bud, a tiny patch of clear skin showing a pin-head point, and vaseline smeared around it, on the forehead and ear bases, to prevent any possibility of the disbudding agent running and burning. Special proprietary disbudding " sticks " are sold for the purpose, and should be used with care, the fingers being protected with brown paper.

35

The kid held firmly between the knees, the stick should be damped on moistened blotting paper and carefully dabbed on the bud and immediately around it in a half-penny-size circle. Repeat the dabbing for about a minute, occasionally moistening the stick.

Having treated both buds thus, rest the kid, then treat each bud again until it is rather black, blistery, and soft. As the bud becomes softer, use greater care not to break the skin. Give a final light dabbing after another interval, and finish by gently pressing a small wad of cotton-wool on each place to absorb surplus moisture. Afterwards place the kid by itself and see that it does not rub its head, when it will quickly forget the operation. Make sure also that the mother cannot get at the kid to lick its head.

TYPICAL DAIRY GOATS.

Note the fine, clean appearance, and the splendid udders, of these British Saanens, specially bred for milk production.

MILKING TIME.

It is not difficult to milk a goat, and even the children can give a hand with the family goat. This little girl is quite an expert.

CHAPTER IX.

Milk-O !

A BIG problem to the beginner is that of actually milking a goat, if no such thing has ever been attempted before. Here again you will find your worries disappearing as you get down to actual practice. It is an "art" readily acquired, and even children make good goat-milkers.

And what a satisfying business it is, this particular business of being your own dairyman. Off you go in the early morning. . . . There is an eager " Ba-aa! " in greeting as you approach nanny. Squish, squish . . . squish, squish . . . the warm milk frothing in the pail, gradually mounting—one, two, three, "Ah! three and a-half pints this morning, my dear ! I bet that's one up on old Smith. Hope I meet him in the train!"

Some people squat on their heels to milk. You may find it more to your convenience to have a low bench on which the goat will quickly learn to stand, while you sit at the side on a stool. A suitable bench would be from 12-18ins. from the ground, the width 2ft., and the length approximately 3½-4ft.

A small bucket may be used for milking. Personally, I frequently use a 3-4 pint enamel mug, over the top of which I clip a thickness or two of butter-muslin, with a rubber band, as an additional aid to cleanliness.

The dairy business end of the goat is a delicate piece of machinery which must be treated as such. The udder must not be tugged or swung about or carelessly knocked. Treat it with firm gentleness.

The mechanics of milking are these: Grasp the further teat with the right hand, in the crutch between the thumb and forefinger. Squeeze, and with reasonable pressure follow up with the second and third fingers. If the teat is long enough it may be possible to use the little finger also. Relax the teat and begin again with the forefinger.

37

Perform the same operation on the other teat with the left hand, alternating with the right.

Note that there is no pulling or stretching of the teat, but only pressure. Occasionally with your wrist you should "bunt" upwards into the udder, in natural imitation of the bunting action of a kid when suckling. This induces a release of milk from the udder.

When you have extracted as much milk as possible in this way, you must "strip" the bag. This is most important. With each hand alternately, draw thumb and forefinger quickly up and down each teat, until the last drops of milk have been extracted. You need these "strippings" as they are the richest part of the milk. Gentle massage of the udder, particularly round the back, during this process, will be found to bring down more milk. Unless an udder is stripped right out at each milking the yield will gradually decline, so do not skip this part of the job.

Speed comes with practice, but always milk as quickly as possible. It ensures a steadier and increased flow, and avoids irritating the goat.

A goat should chew her cud contentedly during milking. If she does not, or is very fidgetty, endeavour to find out what is wrong. You may be pulling long hairs on or about the udder; or there may be a sore or chap on the teat.

For the first few days after kidding milk your goat particularly carefully and lightly; only take a little milk, but gradually increase the amount taken away, until you are milking "full out."

If your goat is a first-kidder she may give you a little trouble until she has become accustomed to an experience which is as strange to her as it is to you. You must be patient, petting and coaxing her, and treating the udder with care.

Very often the teats of a first-kidder are so small that they cannot be grasped easily with the hand. These can be gradually brought into better shape by gentle pulling and massage after applying lanoline or one of the proprietary udder salves. Persistent massage has worked wonders with many an unpromising udder.

CHAPTER X.

Making Your Own Butter and Cheese.

Not least of the joys of keeping goats is that you can make you own butter and cheese. It is surprising how many people think that elaborate equipment is necessary for butter-making. This is far from being the case. Any goat-keeper with a fairly good supply of milk can make delicious butter with the aid of a few simple articles to be found in any kitchen.

I have known butter to be made regularly by being shaken up in a jam jar. I have myself made a weekly quota in an ordinary pudding basin, using a cheap geared cream-whisk. An average of 6 ozs. of butter from just over a gallon of milk has been obtained, with hand skimming. The milk, being skimmed, has still been quite creamy, and served its ordinary purpose in tea, etc.

The pudding basin method is this: After milking, the milk is strained, and cooled as quickly as possible (by standing in a bucket of cold water, if there is no other means), and set aside in shallow pie dishes for from 24 to 36 hours. A secret of obtaining good cream is to see that the dishes stand quite level, little wooden wedges being tucked under them if necessary.

The cream is skimmed off with an enamelled spoon, taking care not to bring away milk with the cream, and put aside in a bowl. Add a pinch of saltpetre and a *good* pinch of salt to every half-cupful of cream. Add salt with each subsequent batch of cream, stirring it in. In about four days sufficient cream will have been obtained to make the butter, and it will be at a good stage of "ripeness." Do not add fresh cream to your supply on the morning of making as this will not be sufficiently ripe.

The butter is made by whipping the cream, slowly at first and working up to a good speed. In a matter of minutes the cream will become granulated. A little cold water must then be poured in and the whipping continued until the butter grains are the size of wheat grains.

39

The buttermilk can now be poured off, to be used for scone making, and more cold water poured into the basin with the butter to " wash " it. Work the butter well with Scotch hands, flat pieces of wood, or wooden spoons previously stood to soak in water and rubbed with salt to prevent the butter sticking. Smack and work the butter until all the buttermilk is washed out, changing the water frequently till it finally comes clear. All that remains then is to beat the butter firmly to shape and extract all moisture possible.

To colour the butter is quite simple. During the whipping process two or three drops of butter-colouring annatto may be added. This can be purchased in small bottles from agricultural chemists. Or you may prefer to adopt the method now used by many goat-keepers of adding one or more egg yolks to the cream before churning. Not only does this give the butter a nice appetising colour but also makes it more nutritious.

Actually it is really no more difficult to make butter in a small table churn, which can be purchased from large stores and dairy supply firms. Naturally, more cream and better butter can be obtained with the aid of a cream separator, but these are a little too expensive for the family goat-keeper to consider, who will have nothing to be ashamed of in the quality of butter she can make in her pudding basin. Let me point out, too, that goats' milk butter has been officially declared an unrationed product, and may be sold freely without coupons.

Goat's Milk Cheese.

There are one or two simple cheeses that the family goat-keeper can make, mainly of the cream or soft variety. Apart from these cheese-making requires rather more elaborate equipment than the average family can run to, and rather more attention to detail than could be found time for.

What I call " Family Cheese " is probably the simplest of all goats' milk cheeses to make. To one quart of fresh warm milk add half a teaspoonful of rennet, and stir well. Let stand for 12 hours, then cut the curd into pieces of uniform walnut size with a knife or enamelled ladle.

Transfer the cut curd to cheese cloth and hang up to drain for 24 hours. Scrape the sides to assist drainage.

Then take down, flavour with a little salt (it is a unique idea to use celery salt), working it in and pressing to shape by standing in a cup or bowl with a weight of some sort on top. Or use one of the proper little moulds which can be purchased from dairy supply shops. The mould should be stood on a small straw mat. Place another mat on top and slightly weight down. Keep thus for two days, turning the mould over and weighting from the other side once. The cheese can then be used.

Cottage Cheese is a well-proved recipe. This is made from skimmed milk, and if you have been making butter you will find it a profitable use for the milk left from the cream setting.

Bring the milk to a temperature of 75 deg. Fahr., and keep it there. Add " starter " or a little buttermilk to speed up the process of curdling, which should not take more than 24 hours. Cut the curd into 2in. cubes with an enamelled ladle and transfer it to a vessel submerged in another vessel of hot water. Raise the temperature to 100 deg. Fahr. " Cook " for 30 minutes, and stir at intervals of five minutes. When ready, place the curd on a cheese cloth to drain, and after ten minutes tie up in the cloth and hang it up to drain.

When the whey stops running out, take down the cheese and well work it with clean boards or Scotch hands, adding salt at the rate of 2½oz. to 10lb. of curd. It is not necessary to use rennet for this cheese, but some makers do so, and claim that a considerable flavour is imparted by it in time.

A simple cream cheese which can be made without rennet is this : Take some thick cream and cool it to 65 deg. Fahr., in three hours. Then hang it to drain in a draught but in a moderately warm place. Afterwards scrape down, and turn the sides to the middle at intervals of six hours. When nearly set add salt to taste, and mould up for use while still sweet.

For the following cream cheese, made with rennet, you require equal quantities of cream and new milk which should be mixed together thoroughly. Heat in another vessel of hot water, bringing the temperature to 80 deg. Fahr., then add 1 c.cm. of rennet to 1 gallon of mixture. After about an hour cut the curd into 1½in. squares, ladle it into cheese cloths, and drain as above.

A Bondon type cheese can be made, enabling you to use

up buttermilk, or sour milk. A mixture of two pints of either with 1½ gallons of whole milk will make about 12 cheeses in the regulation Bondon moulds.

Raise the temperature of the milk to 70 deg. Fahr., and add rennet at the rate of 1 c.cm. to 1 gallon, leaving to set. If the milk is set at night, the curd will be ready for ladling into a cloth for draining next morning. When the curd is drained change it into fresh cloth and tie up with a few pounds pressure placed on top for about 24 hours. Then mix in a little salt and place in the moulds, leaving the moulds to stand on straw mats until properly drained.

It may help you to know that about 17 drops of rennet equal 1 cubic centimetre (c.cm.).

CORNISH SCALDED CREAM.

It will probably interest you to know how to make this, and the following excellent recipe was sent me by a lady in Cornwall: —

Put three or four inches of milk in an enamelled bowl, and leave about 12-24 hours, according to whether the maximum yield of cream is needed or the scalded milk wanted for use as soon as possible. Put the bowl over *gentle* heat. When a thick, wrinkled crust has formed and begun to crack, put the bowl in a cool place for about 24 hours. The cream can then be taken off the milk with a skimmer previously dipped in cold water. The scalded milk makes good puddings.

CHAPTER XI.

Goat Meat for the Table.

A MAN who is a butcher as well as a goat-keeper is perhaps better qualified to deal with this subject than many others. Mr. E. T. Harding is both, and this is how he writes to me: —

" I would not be without my goats for all the world. I have five healthy children who have been reared on goats' milk and who are admitted by all who have seen them to be superior, in build, to most others of similar age.

" The present war situation has proved to most people the advantage of keeping goats for their milk supply. We are told by the authorities not to waste a single thing —so why waste unwanted male kids ? These can be used for the table at any age from two weeks. In Ireland they are considered very good at that age. However, they are at their best when four to five months old.

" The quality depends on how carefully they have been reared. If this has been given consideration the meat, when dressed, will look much like lamb, but fatter inside as is the case with goats. The fat is white, and the liver, heart, etc., will be found to be of exceptionally healthy condition and delicious eating.

" The slaughter, from a humane viewpoint, is best performed by a butcher who will, no doubt, be willing to oblige. In addition, he would probably split the carcase, from the tail to the neck and, placing a knife in the tenth rib from the neck, quarter off. The leg is severed from the loin, these forming roasting joints in addition to the shoulder, which is separated from the neck by a vellum.

" The remainder of the carcase makes boiling meat. The carcase should be hung at least a week before it is cut up or it will tend to prove ' rubbery,' as would a fresh killed lamb.

" The kids are best killed before the mating season, as otherwise the meat may be tainted by odour."

43

I can endorse every word Mr. Harding says, having eaten many a home-reared kid. My local butcher slaughters and dresses the carcase at little more than eighteenpence a time.

"GACON."

Then there is " gacon "—or goat meat bacon. Another goat-keeper has written to me as follows : —

" Last autumn I had a very fat 18 months old kid that would not mate, so I got the butcher to slaughter and dress it. My wife cured it as she does bacon, and it is as sweet and tasty as anyone could want. The hams and shoulder pieces we boil, and fry the sides as bacon.

" I am never going to feed an expensive pig again, but always have a winter's supply of gacon. A castrated billy of a large breed at 18 months need not cost much to feed, and makes a good-sized carcase.

" There are probably several ways of curing a carcase, but this is my wife's method:

" Remove the neck and split the carcase down the middle. Cut off the feet and cut out the ham. For curing use a stone slab if possible. For every 20lb. of meat take 1½lb. of salt, 1oz. of saltpetre, and ½oz. Demarara sugar.

" Rub the meat with only the salt and leave with the skin side downward until the following day. Throw away all liquid that has drained out, until the meat stops discharging a day or two. Then apply the saltpetre to the leaner parts, rubbing it in ; also rub in to the whole the mixture of salt and sugar, turning the meat each day and making the mixture last several days.

" When all the mixture is used the meat need not be turned more than once a week. Collect the drainings each day and pour over the meat. Allow the sides and hams to soak 21 days.

" After curing, soak the gacon in chilled water for 15 minutes, then dry thoroughly with a cloth and hang in a draught to dry. When dry, hang in a room not too hot or the gacon will dry up too much, and one not too damp, either.

" Then you can look forward to many enjoyable rashers."

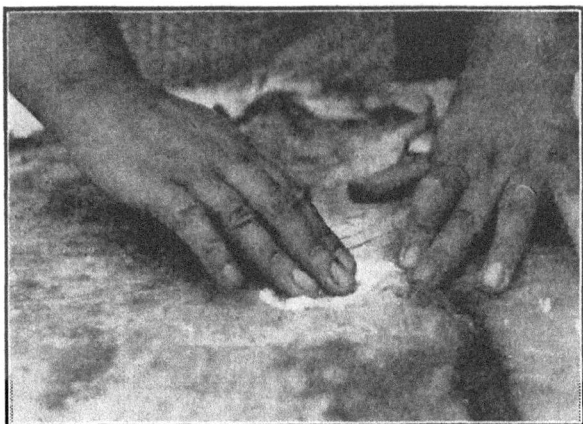

The skins of unwanted kids can be made up into many beautiful articles.
The curing process is simple, the skin being treated with a mixture of
alum, salt and saltpetre.

After treatment the skin is wrapped tightly in newspaper and put away
for ten days or so.   For full curing instructions see page 45.

CHAPTER XII.

Curing Kid Skins.

I HAVE known goat-keepers who, in disposing of unwanted kids, have disposed also of the skins. This may be an excusable extravagance in peace—at the present time, with the national call to economise, it must not be countenanced.

Many beautiful articles can be made from kid skins, and it is perhaps needless to remind lady goat-keepers of the fur coats, jackets, capes, neckpieces, gloves, baby carriage covers, rugs, and so on, which can be made up.

There are several methods of curing skins, but I learned the following simple formula some years ago, and it has always proved successful: —

The skin to be treated must be as free from fat as possible. Use a blunt table knife to scrape off any that is adhering, the skin being placed fur down on several thicknesses of paper.

The curing mixture is equal proportions by measure of salt, saltpetre, and powdered alum, say about a tablespoonful of each for one or two small skins.

Rub this mixture well into the fleshy side of the skin, using the balls of the three middle fingers. Take care in going round the edges. Should the mixture get into the fur it will cause sweating.

Having been all over the skin, place three or four folds of paper on top, roll up tightly, and put away the package for 10-14 days. Then take out the skin and, with the aid of a blunt knife, peel off the papery tissue. This should come away quite easily, exposing a soft, pliable chamois-like skin.

Another method is to tack the skin to a board, fur down, and paint it once every day for a week with alum solution, finally rub well with emery paper, and then work the skin vigorously between the hands. The skin is then ready for use.

In making up, skins should never be cut with scissors,

or the fur comes out. With the skin laid fur down on a flat surface, and the pattern marked out with a pencil, cut fairly lightly with a sharp penknife. If you cut heavily you will go through the fur. I understand that No. 30 sewing cotton is the best for kid skins. Edges should be over-sewn carefully, too deep stitches or pulling tightly on the cotton should be avoided. Any flaws or thin places in the skin can be cut out and a good piece of fur sewn in.

If readers wish, I can advise them of a firm which specializes in the curing of goat-skins at reasonable prices.

CHAPTER XIII.

Ailments.

OF all domestic animals the goat is perhaps the least liable to disease, and some of the worst of those that attack others leave the goat unaffected. It is, of course, liable to certain ailments, and is more susceptible to trouble in certain circumstances, such as inter-breeding, coddling, mis-managing in regard to rearing, feeding, and so on. But ordinarily, a constitutionally sound goat properly managed will give its owner few worries. It is said that goats in Britain are becoming less hardy, but this I contend is a question of upbringing. Many caprine ailments can well be treated at home, and the goat-keeper should deal with minor illness himself when he can. It is no disgrace, however, but a duty, to seek skilled veterinary attention—blessed phrase!—when required.

Some of the commoner complaints are here dealt with: —

Accidents.—These will happen, even in the best regulated goat herd. Common sense is the best prevention. For instance, do not provide hay in nets for kids; they may become strangled by the netting. Do not tether a goat near a retaining wall with a drop the other side. 'Ware barbed wire. Keep horned and hornless goats apart.

Every goat owner must keep a small medicine chest handy, and know how to use the contents to best advantage. A knowledge of simple first-aid can be acquired readily these days !

Blood-in-the-Milk.—When milk is found to be streaked with blood, this may be the result of a chill, blow, or strain causing rupture of some of the small blood vessels in the udder, or, according to one authority, too rich feeding causing over-stimulation of the mammary glands. Gentle milking, a dose of Epsom salts (1 tablespoonful to a $\frac{1}{4}$ pint of water), the temporary cutting down of the rich ingredients of the ration, a bran mash every other

47

day, and liberal green food will speedily cure. I have frequently found that this condition indicates the need for toning up the whole system by giving a course of iron tonic.

Chills and Colds.—A drink of half a pint of warm beer, or a quinine tablet, a sack or rug over the body, and a spell in a draught-proof stable will nip many a chill in the bud. The symptoms develop in the obvious way—a staring coat, shivering, hunching of the back, and later the usual runny nose and lost appetite. Proprietary medicines are available for treatment in advanced stages.

Constipation and Indigestion.—A good milking goat is a big eater and is therefore more prone to digestive disorders, and should have treatment at once before it becomes serious. If her appetite is not quite so keen, or she has "lumpy" droppings, something is wrong. I find the following treatment mostly puts matters right:

Get one pint of pure linseed oil from the chemist and give her two tablespoonfuls of this daily for, say, a week, or until she is normal again. If "blown" through gorging on wet greenfood, a teaspoonful of pure turpentine can be added to the oil. During treatment give plenty of good hay and water.

Diarrhoea or Scour.—This distressing trouble is usually a symptom of some other disorder, such as chills, worms, wrong feeding, etc., and the cause must, of course, be searched for in order to ensure permanent cure. Kids may scour because of overfeeding, or receiving milk at too high or too low a temperature, or lack of cleanliness. There are several remedies: Arrowroot biscuit; a teaspoonful of prepared chalk in milk; white of egg in a little water; dry flour baked in a pan until browned, mixed to a thin paste in warm milk with a little starch added to the first dose, and fed to the kid a little at a time at frequent intervals.

With older animals, all that may be necessary to remove the irritant matter is a dose of castor oil, varying according to age from a teaspoonful to four ounces, following with a little prepared chalk (up to a tablespoonful for an adult) or arrowroot biscuit. If the diarrhoea persists, give from a small teaspoonful to a dessertspoonful of the following mixture every eight hours:

Compound tincture of morphia and chloroform, 4

drachms; liquid bismuth, 4 drachms; oil of cloves, 1 drachm; cooled linseed tea, 7 ounces.

Linseed tea is made by putting ¼lb. of whole linseed in a saucepan with half gallon of water and boiling very slowly on side of stove all day, then straining.

Veterinary chemists, of course, keep prescriptions to treat diarrhoea that they can make up for your goat.

If diarrhoea is due to worms, treat for worms as described on page 50.

Foot Trouble.—If your goat goes lame it may be because the hooves want trimming, or a thorn or stone is causing discomfort. But if the feet are hot and tender it may well point to your goat having too rich a ration and too little green food. An old-fashioned remedy for lameness of this type is a mixture of lard and sugar. Bran and linseed meal poultices are another form of treatment. At the same time, the diet must be cooling, and a dose of linseed oil given daily.

Flies.—When these are a worry to goats, one of the many proprietary fly dressings may be applied along the back. Oil of citronella, oil of lavender, and a wash made by macerating walnut leaves in vinegar, are also useful.

Garget, Caked Bag, or *Mammitis,* as it is variously called, is the principal udder trouble you must guard against, and what it really amounts to is inflammation of the udder. The causes are many, but chiefly ill-health, or lying about on cold, damp ground or wet grass, or injury to the udder.

The first sign is the hardening of the udder, which may become quite lumpy, and curdly milk drawn. A proprietary "udder drench" should be given the goat and she should be placed in a warm, dry, well-bedded loose-box. Hot fomentations should be applied to the udder, which should be particularly carefully dried afterwards so that there is no risk run of a further chill. The udder should also be well massaged, and all the milk or curd that can be extracted should be. The udder should then be well rubbed with one of the recognised udder salves, or an ointment made of camphor mixed with lard. This treatment should be repeated every two or three hours until the udder returns to normal.

Treated promptly in the early stages, the goat will soon be as right as rain again, but if neglected you stand the risk of losing your goat.

E

Lice.—These are not likely to trouble the family goat kept well-groomed and cleanly, but should they appear, tobacco powder, or one of the proprietary lice powders, should be dusted into the coat the wrong way of the hair.

Sore Teats.—Hard, cracked or chapped teats will be quickly soothed and cured by the application of udder salve. A tin of one of the many preparations available should always be kept handy. You need not fear that they will taint the milk. A soothing lotion which can be made up at home, if preferred, consists of boracic acid, 10 per cent. in solution ; a similar quantity of glycerine, and water. Put in a wide-mouthed jar and dip the teats into it after each milking.

Tainted Milk.—The most frequent causes of unpleasant flavour in milk are that the goat is in poor condition, is suffering from indigestion or worms, " strong " foods such as turnip or cabbage have been fed just before milking, or insufficient care has been taken over cleaning milk utensils. These suggest their own remedies. In the latter case one is particularly liable to get " almond-tasting " milk, and I cannot over-emphasise the importance of thoroughly scrubbing all utensils with *cold* water first before they are scalded out with hot water. Even boiling water will not cleanse milky things unless cold water is used first. As an experienced goat-keeper once said to me : " I found, to my surprise, that washing up dairy utensils was a real ' job ' to learn."

Worms.—A goat with worms will develop poor condition, her breath will be offensive, and her coat staring. There may be recurring diarrhoea, and worms may actually be seen. There are several excellent goat worm cures on the market, and you will find it quite simple to dose your goat and get rid of the parasites. If a goat is wormed regularly in spring and autumn there will be little danger of the parasites gaining a hold unsuspectingly and pulling the animal down in condition. The following is my own method :

I use a well-known make of worm pills, giving the goat two at a time (comprising one dose), late in the evening. I give a light feed of hay only about 8 p.m., following with the pills at 10 p.m., or later if possible, and giving no further food till morning. Sometimes the pills can be hidden in stale bread, and the goat takes them well. But the best method is to straddle the animal (as though riding horseback), raise her head up and back with your left hand, force jaws open gently with your right hand, push pill well down the throat, and then, still holding head up, close mouth and nostrils with right hand. The goat is then compelled to swallow. Remember to keep the head up and insert each pill separately. This dose can be repeated in about a month's time. Say this is done in the early autumn, it should be unnecessary to worm again until the spring, or six weeks after kidding. Never worm an in-kid goat.

The worms may be expelled from 12 to 36 hours following dosing. If the goat is stall-fed or in her stable at the time, be sure to remove *all* droppings and burn them.

After worming, a course of tonic will put the goat in fine fettle again. For tonics, see page 54.

CHAPTER XIV.

In General.

Your goat's hooves need trimming regularly, about once a month right from her kid-hood, to keep her in good health, as to let the horn grow so long that it bends under the foot and the edges turn into the sole makes her incapable of walking, and soon pulls her down in condition. She is unable to feed or exercise properly, which causes the milk yield to drop, and then, needless to say, it is unprofitable for you.

As a beginner you may feel incapable of doing the hoof trimming yourself. Perhaps you have a blacksmith near you, or an animal centre or a goat-keeper, who would do the job for you. But really, it is not a difficult job especially if you do it regularly, as then only a "trim up" is required each time.

You should examine the feet every three or four weeks and, if possible, take your goats out for a walk on hard surfaces so that the horn will naturally get worn down.

By far the best tool for a "hoof cutting operation" is a pair of sharp-edged pincers. Gradually clip through the horn with these, as you would with your ordinary nail clippers. A razor-sharp knife is also recommended for this job, but I think for a beginner the first way is the best, as there is less likelihood of doing harm to the goat.

If the horn is very hard, give your goat a run in the wet grass. Or rub oil into the hoof several times previous to the trimming.

Before cutting get some water with a drop or two of disinfectant in it and thoroughly wash the hoof clean, scrubbing it with a brush and taking special care to clean the soles and inner surfaces of the divisions. By doing all this you will be able to see if there are any splinters, cuts, inflammation or soreness in the foot.

If there is much hoof to remove, pare down a little each week, and take the goat on hard ground daily.

Clip or shave the horn down gradually, and never make one deep cut. Shave the point of the toe as well. The

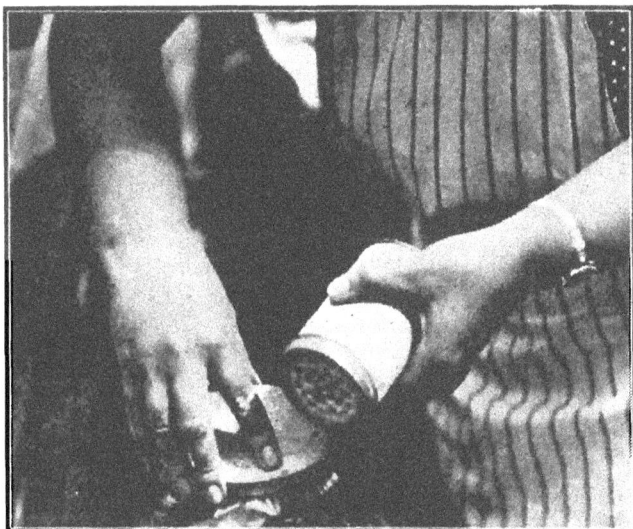

An occasional dusting of proprietary powder, when you groom your goat, will keep her free from undesirable insects. See page 50.

KEEPING THE FAMILY GOAT CLEAN AND SMART

Trimming the hoofs is a routine task the family goat owner must not neglect, to keep the goat comfortable on her feet. How to do it is explained on page 52.

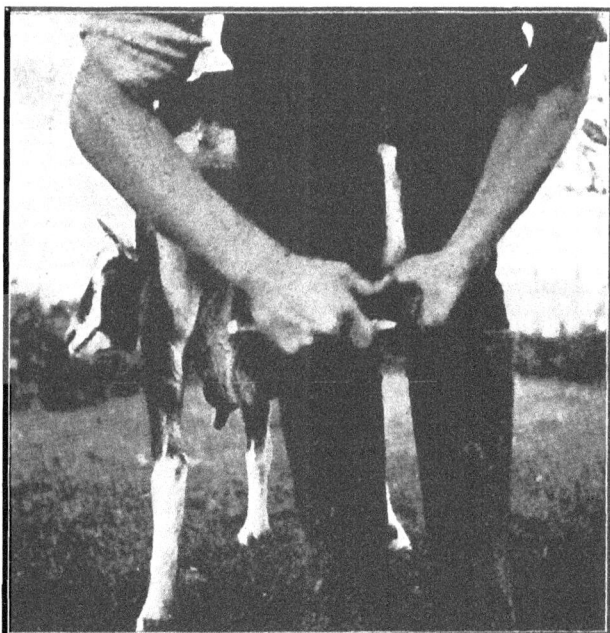

length of a well-shaped hoof is about 2¼in., the depth at the front 1½in., to 1¼in. at the back. Special care must be taken not to draw blood, especially when, if necessary, the soft pad of the heel is cut level with the sole and horn wall. It will be quite all right, so long as you don't cut too deeply.

DETERMINING A GOAT'S AGE.

A goat's age can be fairly accurately estimated by its teeth. Like sheep, goats have no incisors on the upper jaw, but a hard pad of gristle. At one year the full complement of teeth comprises eight small and sharp front teeth or incisors on the lower jaw, and six molars on either side of each jaw, thirty-two in all. The lower front teeth determine the age. At about one year the centre pair of small teeth drop out and are replaced by two larger permanent teeth. Between the twentieth and twenty-fourth month two more large permanent teeth appear, one on each side of the first pair. In the third year two more appear, and in the fourth to fifth year the permanent tooth at each corner comes in. After the fifth year look to the condition of all the teeth to determine age. The more worn the molars, the older the animal. At seven or eight years the front teeth begin to wear or break, and are not replaced. The above " rules " are not infallible, as goats vary considerably in growth according to management, rearing conditions, etc.

GOAT-KEEPING TERMS.

Kid—An animal up to one year old.

Goatling—Over one year but under two years.

Nanny—Female.   *Billy*—Male.

Dam—Female parent.   *Sire*—Male parent.

* *and* Q*—A goat obtaining a minimum number of points in official milking tests is awarded a *. Her daughter, if she qualifies, is entitled to two **, and so on for each generation. Thus: Dora* ; New Dora**.

Q* is awarded to a goat which also gives a percentage of 4 per cent. butter-fat, or over, at both morning and evenings milkings, in tests.

†*Male*—A " dagger " is placed before the name of a male goat if his dam and grandam on his paternal side have qualified as " star " or " Q-star " milkers.

By these signs one can tell at a glance the good goats.

CONDITIONERS AND TONICS.

Good tonics and conditioners are supplied for goats by animal medicine manufacturers. In addition may be mentioned : —

Parrish's Food.—From a teaspoonful to a tablespoonful, according to age, may be given twice daily for a few days.

Powder.—The following mixture may be sprinkled in the food: 1 teaspoonful twice a day for a few days ; equal parts of finely-powdered sulphate of iron, gentian, ginger, aniseed, and bicarbonate of soda.

Iron Tonic.—Buy some lump perchloride of iron in a bottle in which it can be kept well corked. Dissolve two walnut-size pieces in 1 quart of water. Give two tablespoonfuls of this liquid (for adults), either as a drench or mixed in bran mash, twice a week.

Clover Tonic.—Fill a good sized pail with clover hay (the more flowers the better), pour on boiling water, stand for 12 hours, and strain. Add warm water and a little salt before giving it to the goats

Ivy.—A bunch of ivy occasionally acts as a good liver tonic. You need never worry about goats eating ivy. It is good for them. But the berries, which are poisonous, should be avoided.

TO DRENCH A GOAT.

Back the goat into a corner and straddle it, or get your knee against its side. Raise the head with the left hand under the chin, pressing the thumb into the corner of the mouth behind the front teeth. Insert the bottle at the side, keep the animal's neck straight, and touch the roof of the mouth with the bottle, letting the contents slowly trickle down the throat. Rather waste the fluid than choke the animal by pouring too fast.

DRYING OFF A GOAT.

A goat in-milk and due to kid must, if possible, have a rest from milk production to avoid undue strain on her system. A resting period of six weeks is advisable. If your goat does not dry off naturally, you must take steps to dry her off. Well-bred and heavy milkers may show considerable reluctance in this respect—and incidentally, it is often found that those goats which are only going to produce one kid at the parturition are those least inclined to dry off.

Drying off is accomplished by not stripping out after milking, and then, after a few days, skipping every other milking. It should not be done by suddenly taking a little milk only from the udder; this may cause curdling of the milk and caked bag, or painful distention of the udder may start the goat self-sucking. Neither should it be done by cutting down rations. This would have an adverse effect upon the goat's condition.

If the goat persists in milking, don't worry, but feed her extra well to help combat the extra strain.

GREEN FOOD TO GROW.

Milk production costs are lowered in proportion to the use made of home-produced feeding-stuffs. Do not be content to allow your goats only part of, or the " waste " from, your normal kitchen garden. Devote a special plot to goat green food. If your garden is not big enough for the purpose, take an allotment.

A 10-rod plot will supply one goat's green food requirements for the greater part, if not most, of the year. Many crops can be grown in such an area, and the greater the variety the better. A suggestion in this respect is that the plot should be quartered, one quarter being devoted to, say, succession sowings of mustard; another to roots, swedes, carrots, turnips; a third to kale, marrow-stemmed and perpetual sprouting; and the remainder to chicory, which provides a wealth of cut-and-come-again foliage. Sunflowers and artichokes could be planted along the borders, both providing most useful goat food.

Make no mistake, there are very big possibilities in a " goat plot."

CHAPTER XV.

A Few Recipes.

GOAT-MILK BREAD.

The following bread is quickly and easily made. It can be kept several days, or it may be eaten hot without fear of indigestion: —

Required: ½lb. plain flour, 1 small teaspoonful bicarbonate of soda, 1 small teaspoonful cream of tartar, 1 breakfast-cupful of goat milk.

Mix the dry ingredients; add milk to make a dough. Put in buttered tin and place in hot oven. Let oven cool slightly and cook for about half-hour, testing with a skewer.

GOAT-MILK CANDY.

Required: 1 lb. sugar, ¼ pint goat milk, 2 ozs. 'margarine, ½lb. coconut.

Place in a saucepan; boil fairly slowly and stir occasionally to prevent burning. When the mixture bubbles stir in the coconut. Then turn into flat tins or plates, and cut into squares when quite cold.

CHOCOLATE CUP.

Take 1 pint goats' milk, 1 tablespoonful cocoa, 1 tablespoonful sugar. Mix cocoa and sugar together. Add 1 pint of cold goats' milk, stirring well. Then pour in 1 pint of boiling water, stirring all the time.

CAPRA CUSTARD.

Required: Three or four bananas, 1 tablespoonful of sugar, 2 eggs, ¾ pint of goat milk.

Beat the eggs and sugar together in a basin and add boiling milk. Butter a pie-dish and pour this custard in, adding the bananas, which should have been peeled and thinly sliced.

Place the dish in a moderate oven and bake until the custard has set.

GINGER GOAT CAKE.

Required : 8ozs. self-raising flour, 4ozs. margarine or goat butter, 4ozs. sugar, 2 teaspoonfuls ground ginger, 1 egg, a little goat milk, and lemon flavouring if desired.

Rub the butter into the flour, mix in sugar, ginger, and flavouring, beat in the egg and milk. Put the mixture into a paper-lined tin and bake in a moderate oven.

TEA SCONES.

Use soured goat milk to make these tea scones:

Take 6ozs. self-raising flour, 3 ozs. butter or margarine, 1 teaspoonful castor sugar, a few sultanas, a pinch of salt.

Mix with sufficient sour goat milk to make into a dough suitable to roll out. Shape and cook 10-15 minutes in a hot oven.

NIGHT CAP FOR ONE.

Take one dessertspoonful of oatmeal, a teaspoonful of syrup, a pinch of salt, cold water, and ½ pint goats' milk. Put oatmeal in a pudding basin and just cover with cold water. Heat the goats' milk to boiling point and pour it into the bowl; add pinch of salt and stir in the syrup. Allow the oatmeal to settle and then drain off the liquid contents into a glass.

CHAPTER XVI.

A Family Man Starts Goat-Keeping.

A FITTING conclusion to a book on family goat-keeping is the following reprint from GOATS Magazine—August, 1940—in which Mr. H. V. Cook, a typical beginner, describes his experiences with his first goat. Anyone who is still hesitating whether or not to take up goat-keeping will find much encouragement in this frank and detailed account.

It was about February, 1940, that the hunch of obtaining a goat first came into the mind of the writer—the main idea being to assure the family of a fresh supply of milk each day, particularly for our boy then six months old; the idea was important in view of the war situation.

However, there was no intention of just going out and buying a goat at random and hoping for the best—my intention firstly was to acquire more than the scant knowledge of goats which I had at that time.

The idea was then mentioned to several people and was met with wry smiles and the most amazing revelations about these four-footed animals. In fact I was astounded to find so many people "knew" all about goats without ever having seen one, let alone kept one. According to these experts goats were useless, they smelt horribly, I would never be able to acquire the taste for their milk since it was so strong and tainted, they would eat everything around the place, and were often ferocious, etc., etc. These were a few of a barrage of detrimental remarks pertaining to goats.

Having obtained so much information I must admit that my "hunch" somewhat abated, until one day I came across an advertisement for "GOATS MAGAZINE," and without delay approached our Editor for advice—the reply was very different from my previous informers—with the result that I hit upon the idea of going out and actually finding someone who kept goats in order to make a more exact enquiry before taking the plunge and obtaining one.

This idea was really the best that had been hit upon, since in my travels around the country I was able to inspect some lovely herds, picked up a lot of very useful hints, and, most important of all, saw for myself how goats should and should not be cared for—the contrasts were so obvious to me that I came to the conclusion right away that a lot of people keeping goats were not getting the best out of them.

This more detailed survey convinced me upon the matter, so it was decided to obtain a goat in-milk without delay. At this time I met my first obstacle. Much to my dismay it was easier said than done—at every turn I was met with the same tale: nothing to sell now, but there will be some kids available soon.

Once again I approached our Editor, and after some weeks he was able to put me into touch with the party from whom I bought a small white hornless nanny with two nanny kids.

THE ARRIVAL.

It was a Saturday in April when they arrived at the station—the kids in a box, and nanny with a sackcloth cover over her. They caused a considerable amount of excitement when I finally persuaded her to come out of the van on to the platform. She appeared to be very distressed and nervous and just refused to move an inch despite all kinds of coaxing. Finally I hit upon the idea of taking the kids along; she then followed, at the same time telling all and sundry that those were her kids. I suppose the sight was a little unusual as we all went down the platform, and I did feel a little red about the neck. At last they were coaxed into my car ready for home.

The "hunch" was now a thing of the past, and upon arrival home the kids were firstly unpacked and the cover removed off nanny. Naturally all arrangements, such as accommodation, food, and so forth, had been seen to previously, and in a few minutes they were all taking stock of the new home. Food and water were ignored, so they were just left to their own devices for the rest of that day in order to let them settle—even the Dalmatian dog was locked away out of sight since upon getting a glimpse of him the nanny got excited and upset.

Information was supplied with this little family stating all particulars about food, water and milking; also the fact that the kids be taken away from her each night and then allowed to be with her during the day, after about one quart of milk had been drawn off her each morning. This latter idea suited me very nicely owing to the varying hours at which I arrive home from business at night, in addition to saving my wife the trouble of bottle feeding the kids during the day, her hands being more than full looking after a couple of kids of the two-legged variety, to say nothing of other domestic requirements.

I made the first attempt at milking nanny on the Monday morning, tying her head up close to a staple in the wall (as per instructions). I would mention that the only experience I had in milking was many years ago as a boy on the farm where I often helped to milk the cows. Needless to say I found very quickly that trying to milk a "first-kidder nanny" was different. I was prepared for anything to happen as I took my seat at the business end. The smallness of the teats was my greatest handicap; after some time I did manage to draw off about half a pint, and feeling that the journey and new surroundings, coupled with the fact that so far she had eaten little and drank less, would have all tended to have upset the supply of milk, I gave it best for that day and put the kids back with her, knowing full well that they would take care of what was left.

SETTLING DOWN.

Each morning afterwards I gradually increased the amount drawn off until I was getting the quart. She was eating and drinking well at the end of a week and was more or less settled, although she still did not like the dog anywhere in sight. On the fifth morning I milked her without tying her head, and she behaved perfectly.

All went well until the kids were seven weeks old, when it was noticed that the smaller of the two was showing signs of weakness on the rear legs. I inspected her very carefully, thinking that she might have twisted herself romping about; there was no apparent pain or soreness so I just decided to watch carefully. During the following days the trouble got worse; in fact she could only walk a few yards after which she had to lie down and rest; furthermore she appeared thin and scraggy. Drastic action was needed, so without delay I started her on a daily dose of cod-liver oil and took precautions to see that she got more.milk; this treatment did the trick, for at the end of a week there was a marked improvement; now at three months she is cured and there is every chance of her beating her sister for size. I am not dosing the other one since she appears to be full of life and I am keen to see what the final outcome of this treatment is.

They are turned out in the field all day when the weather is fine—when it's wet I keep them in, but despite ample feeding there is a marked reduction in milk yield next morning.

My greatest trouble has been in getting her to drink, some days she will and others she won't. I give her warm water with a little salt in it with a view to encouraging the drinking habit.

Nanny is now quite used to the dog being about; she has grown to know us all, and always greets me when she hears the car come in. The kids are now eating concentrates and hay, both being " full of beans."

During the few months spent with this goat and her kids I have learned much and am still learning. I must say that it is possible to train them to your own requirements; there is no trace of smell provided they are cared for with the small amount of attention that any domesticated animal requires. As far as the milk is concerned I cannot notice any difference to cows' milk. I have tried several friends on with it and so far they have not complained or noticed anything different.

My boy was weaned, put on to a milk substitute and suffered badly with wind; in fact he was troubled with this painful complaint from the very first. When he was put on to goats' milk the trouble went like magic and to say the least, without making any ridiculous claims for goats' milk, he is a picture, and it has certainly done him more good than harm.

Recently it was noticed that the nanny and kids were doing a lot of scratching and upon inspection found that they had parasites in a big way. It was a lovely warm day so a mixture of carbolic and warm water, together with some carbolic soap, was rubbed well into their coats —they all appeared to enjoy the treatment and after a good rub down and brushing appeared to be in a more

contented state. I intend to repeat this treatment at
weekly intervals (weather permitting) for three weeks in
order to take care of eggs hatching out—once they are
clear I see no reason why they should not be kept so
with very little trouble.

Arrangements have been made for winter fodder,
inasmuch as a good sowing of artichokes, swedes and
other suitable winter feed have been made in the garden.

The nanny is not a heavy milker (about four pints)
and it is hoped that the yield will increase upon her
second kidding; in addition one of the kids appears to
have the making of a very fine goat in her, and at three
months old has every sign of being of a heavier milking
strain than her mother.

PROFIT—AND EXPERIENCE.

At this time it can be said that I am nothing out of
pocket, since I still have the stock, worth more now than
when it was bought, and furthermore the experience
gained has been well worth the ten to fifteen minutes
spent daily in cleaning, milking and feeding, and last but
not least that little bit of fuss which they always appear
to appreciate so much.

In the near future my little daughter Jacqueline, aged
four, will be able to take over many of the little jobs,
such as milking and grooming. Already she does much
to help and takes a great interest in them; in fact she has
been itching to have a go at milking nanny from the very
first.

www.ingramcontent.com/pod-product-compliance
Lightning Source LLC
Chambersburg PA
CBHW071750270326
41928CB00013B/2866